To a lover
named Paul

From a lover of Nature
named Paul

THE LIFE OF INLAND WATERS

SPRING
FLOOD:

SUMM
SUNSH.

The view is from West Hill, looking across the he
Field Station towards t

AUTUMN
FIRES

WINTER
FREEZING

of **Cayuga** Lake and the grounds of the Biological
Campus of Cornell University.

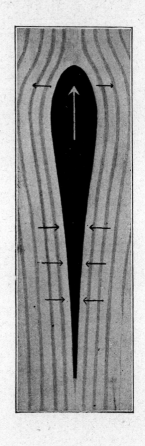

THE LIFE OF INLAND WATERS

An elementary text book of fresh-water
biology for students

By

JAMES G. NEEDHAM
Professor of Limnology in Cornell University

and

J. T. LLOYD
Formerly, Instructor in Limnology in Cornell University

THIRD EDITION

ITHACA   NEW YORK
COMSTOCK PUBLISHING COMPANY, INC.
1937

PRINTED IN THE UNITED STATES OF AMERICA
THE COLLEGIATE PRESS · MENASHA · WISCONSIN

PREFACE

IN THE following pages we have endeavored to present a brief and untechnical account of fresh-water life, its forms, its conditions, its fitnesses, its associations and its economic possibilities. This is a vast subject. No one can have detailed first hand knowledge in any considerable part of it. Hence, even for the elementary treatment here given, we have borrowed freely the results of researches of others. We have selected out of the vast array of material that modern limnological studies have made available that which we deem most significant.

Our interests in water life are manifold. They are in part economic interests, for the water furnishes us food. They are in part aesthetic interests, for aquatic creatures are wonderful to see, and graceful and often very beautiful. They are in part educational interests, for in the water live the more primitive forms of life, the ones that best reveal the course of organic evolution. They are in part sanitary interests; interests in pure water to drink, and in control of water-borne diseases, and of the aquatic organisms that disseminate diseases. They are in part social interests, for clean shores are the chosen places for water sports and for public and private recreation. They are in part civic interests, for the cultivation of water products for human food tends to increase our sustenance, and to diversify our industries. Surely these things justify an earnest effort to make some knowledge of water life available to any one who may desire it.

The present text is mainly made up of the lectures of the senior author. The illustrations, where not otherwise credited, are mainly the work of the junior author. Yet we have worked jointly on every page of the book. We are indebted for helpful suggestions regarding the text to Professors E. M. Chamot, G. C. Embody, A. H. Wright, and to Dr. W. A. Clemens. Miss Olive Tuttle has given much help with the copied figures.

Since 1906, when a course in general limnology was first established at Cornell University, we have been associated in developing an outline of study for general students and a program of practical exercises. The text-book is presented herewith: the practical exercises are incorporated in a small brochure by J. G. and P. R. Needham entitled *Guide to the Study of Fresh-Water Biology*.

The limitations of space have been keenly felt in every chapter; especially in the chapter on aquatic organisms. These are so numerous and so varied that we have had to limit our discussion of them to groups of considerable size. These we have illustrated in the main with photographs of those representatives most commonly met with in the course of our own work. Important groups are, in some cases, hardly more than mentioned; the student will have to go to the reference books cited for further information concerning them. The best single work to be consulted in this connection is the *American Fresh-Water Biology* edited by Ward and Whipple and published by John Wiley and Sons. Our bibliography, necessarily brief, includes chiefly American papers. We have cited but a few comprehensive foreign works; the reference lists in these will give the clue to all the others.

It is the ecologic side of the subject rather than the systematic or morphologic, that we have emphasized. Nowadays there is being put forward a deal of new ecologic terminology for which we have not discovered any good use; hence we have omitted it.

Limnology in America today is in its infancy. The value of its past achievements is just beginning to be appreciated. The benefits to come from a more intensive study of water life are just beginning to be disclosed. That there is widespread interest is already manifest in the large number of biological stations at which limnological work is being done. From these and other kindred laboratories much good will come; much new knowledge of water life, and better application of that knowledge to human welfare.

JAMES G. NEEDHAM
J. T. LLOYD.

CONTENTS

CHAPTER I

Introduction

CHAPTER II

The Nature of Aquatic Environment

CHAPTER III

Types of Aquatic Environment

CHAPTER IV

Aquatic Organisms

CHAPTER I

INTRODUCTION

INDIANS GATHERING WILD RICE, N. MINNESOTA

THE home of primeval man was by the waterside. The springs quenched his thirst. The bays afforded his most dependable supply of animal food. Stream-haunting, furbearing animals furnished his clothing. The rivers were his highways. Water sports were a large part of his recreation; and the glorious beauty of mirroring surfaces and green flower-decked shores were the manna of his simple soul.

The circumstances of modern life have largely removed mankind from the waterside, and common needs have found other sources of supply; but the primeval instincts remain. And where the waters are clean, and shores unspoiled, thither we still go for rest, and refreshment. Where fishes leap and sweet water lilies glisten, where bull frogs boom and swarms of May-flies hover, there we find a life so different from that of our usual surroundings that its contemplation is full of interest. The school boy lies on the brink of a pool, watching the caddisworms haul their lumbering cases about on the bottom, and the planctologist plies his nets, recording each season the wax and wane of generations of aquatic organisms, and both are satisfied observers.

The study of water life, which is today the special province of the science of limnology*, had its beginning in the remote unchronicled past. Limnology is a modern name; but many limnological phenomena were known of old. The congregating of fishes upon their spawning beds, the emergence of swarms of May-flies from the rivers, the cloudlike flight of midges over the marshes, and even the "water bloom" spreading as a filmy mantle of green over the still surface of the lake-- such things could not escape the notice of the most casual observer. Two of the plagues of Egypt were limnological phenomena; the plague of frogs, and the plague of the rivers that were turned to blood.

Such phenomena have always excited great wonderment. And, being little understood, they have given rise to most remarkable superstitions.† Little real

---

Limnos = shore, waterside, and *logos* = a treatise: hydrobiology.

†The folk lore of all races abounds in strange interpretations of the simplest limnological phenomena; bloody water, magic shrouds (stranded "blanket-algæ"), spirits dancing in waterfalls, the "will o' the wisp" (spontaneous combustion of marsh gas), etc. Dr. Thistleton Dyer has summarized the folk lore concerning the last mentioned in *Pop. Sci. Monthly* 19:67, 1881. In Keightly's *Fairy Mythology*, p. 491 will be found a reference to the water and wood maids called Rusalki. "They are of a beautiful form with long green hair: They swing and balance themselves on the branches of trees, bathe in lakes and rivers, play on the surface of the water, and wring their locks on the green mead at the water's edge." On fairies and carp rings see Theodore Gill in *Smithsonian Miscellaneous Collections* 48:203, 1905.

knowledge of many of them was possible so long as the most important things involved in them—often even the causative organisms—could not be seen.   Progress awaited the discovery of the microscope.

The microscope opened a new world of life to human eyes—"the world of the infinitely small things."   It revealed new marvels of beauty everywhere.   It dis-

FIG. 1.   Waterbloom (*Euglena*) on the surface film of the Renwick lagoon at Ithaca.   The clear streak is the wake of a boat just passed.

covered myriads of living things where none had been suspected to exist, and it brought the elements of organic structure and the beginning processes of organic development first within the range of our vision.   And this is not all.   Much that might have been seen with the unaided eye was overlooked until the use of the microscope taught the need of closer looking.   It would be hard to overestimate the stimulating effect of the invention of this precious instrument on all biological sciences.

With such crude instruments as the early microscopists could command they began to explore the world
over again. They looked into the minute structure of
everything—forms of crystals, structure of tissues,
scales of insects, hairs and fibers, and, above all else,
the micro-organisms of the water. These, living in a
transparent medium, needed only to be lifted in a drop
of water to be ready for observation. At once the early
microscopists became most ardent explorers of the
water. They found every ditch and stagnant pool
teeming with forms, new and wonderful and strange.
They often found each drop of water inhabited. They
gained a new conception of the world's fulness of life
and one of the greatest of them Roesel von Rosenhof,
expressed in the title of his book, *"Insekten Belustigung"** the pleasure they all felt in their work. It was
the joy of pioneering. Little wonder that during a
long period of exploration microscopy became an end
in itself. Who that has used a microscope has not been
fascinated on first acquaintance with the dainty elegance and beauty of the desmids, the exquisite sculpturing of diatom shells, the all-revealing transparency of
the daphnias, etc., and who has not thereby gained a
new appreciation of the ancient saying, *Natura maxime
miranda in minimis.*†

Among these pioneers there were great naturalists—
Swammerdam and Leeuwenhoek in Holland, the latter,
the maker of his own lenses; Malpighi and Redi in
Italy; Reaumer and Trembly in France; the above
mentioned, Roesel, a German, who was a painter of
miniatures; and many others. These have left us
faithful records of what they saw, in descriptions and
figures that in many biological fields are of more than
historical importance. These laid the foundations of

*Belustigung = delight.
†Nature is most wonderful in little things.

our knowledge of water life. Chiefly as a result of their
labor there emerged out of this ancient "natural
philosophy" the segregated sciences of zoology and
botany. Our modern conceptions of biology came
later, being based on knowledge which only the per-
fected microscope could reveal.

A long period of pioneer exploration resulted in the
discovery of new forms of aquatic life in amazing
richness and variety. These had to be studied and
classified, segregated into groups and monographed,
and this great survey work occupied the talents of
many gifted botanists and zoologists through two
succeeding centuries—indeed it is not yet completed.
But about two centuries after the construction of the
first microscope, occurred an event of a very different
kind, that was destined to exert a profound influence
throughout the whole range of biology. This was the
publication of Darwin's *Origin of Species*. This book
furnished also a tool, but of another sort—a tool of the
mind. It set forth a theory of evolution, and offered
an explanation of a possible method by which evolution
might come to pass, and backed the explanation with
such abundant and convincing evidence that the
theory could no longer be ignored or scoffed out of
court. It had to be studied. The idea of evolution
carried with it a new conception of the life of the world.
If true it was vastly important. Where should the
evidence for proof or refutation be found? Naturally,
the simpler organisms, of possible ancestral characteristics, were sought out and studied, and these live in the
water. Also the simpler developmental processes, with
all they offer of evidence; and these are found in the
water. Hence the study of water life, especially with
regard to structure and development, received a mighty
impetus from the publication of this epoch-making book.
The half century that has since elapsed has been one of
unparalleled activity in these fields.

Almost simultaneously with the appearance of
Darwin's great work, there occurred another event
which did more perhaps than any other single thing to
bring about the recognition of the limnological part of
the field of biology as one worthy of a separate recognition and a name. This was the discovery of plancton
—that free-floating assemblage of organisms in great
water masses, that is self-sustaining and self-maintaining
and that is independent of the life of the land. Lilje-
borg and Sars found it, by drawing fine nets through
the waters of the Baltic. They found a whole fauna
and flora, mostly microscopic—a well adjusted society
of organisms, with its producing class of synthetic
plant forms and its consuming class of animals; and
among the animals, all the usual social groups, herbivores and carnivores, parasites and scavengers. Later,
this assemblage of minute free-swimming organisms
was named plancton.* After its discovery the seas
could no longer be regarded as "barren wastes of
waters"; for they had been found teeming with life.
This discovery initiated a new line of biological exploration, the survey of the life of the seas. It was simple
matter to draw a fine silk net through the open water
and collect everything contained therein. There are
no obstructions or hiding places, as there are everywhere on land; and the fine opportunity for quantitative as well as qualitative determination of the life of
water areas was quickly grasped. The many expedi-
tions that have been sent out on the seas and lakes of
the world have resulted in our having more accurate
and detailed knowledge of the total life of certain of
these waters than we have, or are likely to be able soon
to acquire, of life on land.

Prominent among the investigators of fresh water life
in America during the nineteenth century were Louis

Planktos = drifting, free floating.

Agassiz, an inspiring teacher, and founder of the first
of our biological field stations; Dr. Joseph Leidy, an
excellent zoologist of Philadelphia, and Alfred C. Stokes
of New Jersey, whose *Aquatic Microscopy* is still a useful handbook for beginners.

Our knowledge of aquatic life has been long accumu-
lating. Those who have contributed have been of very
diverse training and equipment and have employed
very different methods. Fishermen and whalers; col-
lectors and naturalists; zoologists and botanists, with
specialists in many groups; water analysts and sani-
tarians; navigators and surveyors; planktologists and
bacteriologists, and biologists of many names and sorts
and degrees; all have had a share. For the water has
held something of interest for everyone.

Fishing is one of the most ancient of human occupa-
tions; and doubtless the beginning of this science was
made by simple fisher-folk. Not all fishing is, or ever
has been, the catching of fish. The observant fisherman
has ever wished to know more of the ways of nature, and
science takes its origin in the fulfillment of this desire.

The largest and the smallest of organisms live in the
water, and no one was ever equipped, or will ever be
equipped to study any considerable part of them.
Practical difficulties stand in the way. One may not
catch whales and water-fleas with the same tackle, nor
weigh them upon the same balance. Consider the difference in equipment, methods, area covered and numbers caught in a few typical kinds of aquatic collecting:

(1). Whaling involves the coöperative efforts of
many men possessed of a specially equipped vessel. A
single specimen is a good catch and leagues of ocean
may have to be traversed in making it.

(2). Fishing may be done by one person alone,
equipped with a hook and line. An acre of water affords
area enough and ten fishes may be called a good catch.

(3). Collecting the commoner invertebrates, such as water insects, crustaceans and snails involves ordinarily the use of a hand net. A square rod of water is sufficient area to ply it in; a satisfactory catch may be a hundred specimens.

(4). For collecting entomostracans and the larger plancton organisms towing nets of fine silk bolting-cloth are commonly employed. Possibly a cubic meter of water is strained and a good catch of a thousand specimens may result.

(5). The microplancton organisms that slip through the meshes of the finest nets are collected by means of centrifuge and filter. A liter of water is often an ample field for finding ten thousand specimens.

(6). Last and least are the water bacteria, which are gathered by means of cultures. A single drop of water will often furnish a good seeding for a culture plate yielding hundreds of thousands of specimens.

Thus the field of operation varies from a wide sea to a single drop of water and the weapons of chase from a harpoon gun to a sterilized needle. Such divergencies have from the beginning enforced specialization among limnological workers, and different methods of studying the problems of water life have grown up wide apart, and, often, unfortunately, without mutual recognition. The educational, the economic and the sanitary interests of the people in the water have been too often dealt with as though they are wholly unrelated.

The agencies that in America furnish aid and support to investigations in fresh water biology are in the main:

1. Universities which give courses of instruction in limnology and other biological subjects, and some of which maintain field stations or laboratories for investigation of water problems. 2. National, state and municipal boards and surveys, which more or less constantly maintain researches that bear directly upon

their own economic or sanitary problems. 3. Societies, academies, institutes, museums, etc., which
variously provide laboratory facilities or equip expedi-
tions or publish the results of investigations. 4.
Private individuals, who see the need of some special
investigation and devote their means to furthering it.
The Universities and private benefactors do most to
care for the researches in fundamental science. Fish
commissions and sanitary commissions support the
applied science. Governmental and incorporated insti-
tutions assist in various ways and divide the main work
of publishing the results of investigations.

It is pioneer limnological work that these various
agencies are doing; as yet it is all new and uncorre-
lated. It is all done at the instance of some newly
discovered and pressing need. America has quickly
passed from being a wilderness into a state of highly
artificial culture. In its centers of population great
changes of circumstances have come about and new
needs have suddenly arisen. First was felt the failure
of the food supply which natural waters furnished;
and this lack led to the beginning of those limnological
enterprises that are related to scientific fish culture.
Next the supply of pure water for drinking failed in our
great cities; knowledge of water-borne diseases came
to the fore: knowledge of the agency of certain
aquatic insects as carriers of dread diseases came in;
and suddenly there began all those limnological enterprises that are connected with sanitation. Lastly, the
failure of clean pleasure grounds by the water-side,
and of wholesome places of recreation for the whole
people through the wastefulness of our past methods of
exploitation, through stream and lake despoiling, has
led to those broader limnological studies that have to
do with the conservation of our natural resources.

WATER

"O F ALL inorganic substances, acting in their own proper nature, and without assistance or combination, water is the most wonderful. If we think of it as the source of all the changefulness and beauty which we have seen in the clouds; then as the instrument by which the earth we have contemplated was modelled into symmetry, and its crags chiseled into grace; then as, in the form of snow, it robes the mountains it has made, with that transcendent light which we could not have conceived if we had not seen; then as it exists in the foam of the torrent, in the iris which spans it, in the morning mist which rises from it, in the deep crystalline pools which mirror its hanging shore, in the broad lake and glancing river, finally, in that which is to all human minds the best emblem of unwearied, unconquerable power, the wild, various, fantastic, tameless unity of the sea; what shall we compare to this mighty, this universal element, for glory and for beauty? or how shall we follow its eternal cheerfulness of feeling? It is like trying to paint a soul."—RUSKIN.

CHAPTER II

THE NATURE OF AQUATIC ENVIRONMENT

PROPERTIES AND USES

WATER, the one abundant liquid on earth, is, when pure, tasteless, odorless and transparent. Water is a solvent of a great variety of substances, both solid and gaseous. Not only does it dissolve more substances than any other liquid, but, what is more important, it dissolves those substances which are most needed in solution for the maintenance of life. Water is the greatest medium of exchange in the world. It brings down the gases from the atmosphere; it transfers ammonia from the air into the soil for plant food; it leaches out the soluble constituents of the soil; and it acts of itself as a chemical agent in nutrition, and also in those changes of putrefaction and decay that keep the world's available food supply in circulation.

Water is nature's great agency for the application of mechanical energy. It is by means of water that deltas are built and hills eroded. Water is the chief factor in all those eternal operations of flood and floe by which the surface of the continent is shaped.

Transparency.—Water has many properties that fit it for being the abode of organic life. Second only in importance to its power of carrying dissolved food materials is its transparency. It admits the light of the sun; and the primary source of energy for all organic life is the radiant energy of the sun. Green plants use this energy directly; animals get it indirectly with their food. Green plants constitute the producing class of organisms in water as on land. Just in proportion as the sun's rays are excluded, the process of plant assimilation (photosynthesis) is impeded. When we wish to prevent the growth of algae or other green plants in a reservoir or in a spring we cover it to exclude the light. Thus we shut off the power.

Pure water, although transparent, absorbs some of the energy of the sun's rays passed through it, and water containing dissolved and suspended matter (such as are present in all natural water) impedes their passage far more. From which it follows, that the superficial layer of a body of water receives the most light. Penetration into the deeper strata is impeded according to the nature of the water content. Dissolved matters tint the water more or less and give it color. Every one knows that bog waters, for example, are dark. They look like tea, even like very strong tea, and like tea they owe their color to their content of dissolved plant substances, steeped out of the peaty plant remains of the bog.

Suspended matters in the water cause it to be turbid. These may be either silt and refuse, washed in from the land, or minute organisms that have grown up in the water and constitute its normal population. One who has carefully watched almost any of our small northern lakes through the year will have seen that its waters are clearest in February and March, when there is less organic life suspended in them than at other seasons. But it is the suspended inorganic matter that causes the most marked and sudden changes in turbidity—the washings of clay and silt from the hills into a stream; the stirring up of mud from the bottom of a shallow lake with high winds. The difference in clearness of a creek at flood and at low water, or of a pond before and after a storm is often very striking.

Such sudden changes of turbidity occur only in the lesser bodies of water; there is not enough silt in the world to make the oceans turbid.

The clearness of the water determines the depth at which green plants can flourish in it. Hence it is of great importance, and a number of methods have been devised for measuring both color and turbidity. A simple method that was first used for comparing the clearness of the water at different times and places and one that is, for many purposes, adequate, and one that is still used more widely than any other,* consists in the lowering of a white disc into the water and recording the depth at which it disappears from view. The standard disc is 20 cm. in diameter†; it is lowered in a horizontal position during midday light. The depth at which it entirely disappears from view is noted. It is then slowly raised again and the depth at which it reappears is noted. The mean of these two measurements is taken as the depth of its visibility

*Method of Secchi: for other methods, see Whipple's *Microscopy of Drinking Water*, Chap. V.   Steuer's *Planktonkunde*, Chapter III.

†Whipple varied it with black quadrants, like a surveyor's level-rod target and viewed it through a water telescope.

beneath the surface. Such a disc has been found to
disappear at very different depths. Witness the following typical examples:

Pacific Ocean	59	meters
Mediterranean Sea	42	meters
Lake Tahoe	33	meters
Lake Geneva	21	meters
Cayuga Lake	5	meters
Fure Lake (Denmark), Mar	9	meters
Fure Lake (Denmark), Aug.	5	meters
Fure Lake (Denmark), Dec.	7	meters
Spoon River (Ill.) under ice	3.65	meters
Spoon River (Ill.) at flood	.013	meters

It is certain that diffused light penetrates beyond the
depth at which Secchi's disc disappears. In Lake
Geneva, for example, where the limit of visibility is
21m., photographic paper sensitized with silverchloride
ceased to be affected by a 24-hour exposure at a depth
of about 100 meters or when sensitized with iodobromide
of silver, at a depth about twice as great. Below this
depth the darkness appears to be absolute. Indeed it
is deep darkness for the greater part of this depth, 90
meters being set down as the limit of "diffused light."
How far down the light is sufficient to be effective in
photosynthesis is not known, but studies of the distribution in depth of fresh water algae have shown them
to be chiefly confined, even in clear lakes, to the uppermost 20 meters of the water. Ward ('95) found 64
per cent of the plancton of Lake Michigan in the uppermost two meters of water, and Reighard ('94) found
similar conditions in Lake St. Clair. Since the intensity of the light decreases rapidly with the increase in
depth it is evident that only those plants near the surface of the water receive an amount of light comparable
with that which exposed land plants receive. Less than
this seems to be needed by most free swimming algae,

since they are often found in greatest number in open waters some five to fifteen meters below the surface. Some algae are found at all depths, even in total darkness on the bottom; notably diatoms, whose heavy silicious shells cause them to sink in times of prolonged calm, but these are probably inactive or dying individuals. There are some animals, however, normally dwelling in the depths of the water, living there upon

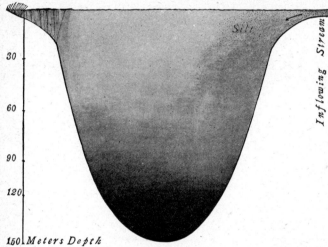

Fig. 3.   Diagram illustrating the penetration of light into the water of a lake; also, its occlusion by inflowing silt and by growths of plants on the surface.

the organic products produced in the zone of photosynthesis above and bestowed upon them in a considerable measure by gravity.   To the consideration of these we will return in a later chapter.

The accompanying diagram graphically illustrates the light relations in a lake.   The deeper it is the greater its mass of unlighted and, therefore, unproductive water, and the larger it be, the less likely is its upper stratum to be invaded by obscuring silt and water weeds.

Mobility—Water is the most mobile of substances, yet it is not without internal friction. Like molasses, it stiffens with cooling to a degree that affects the flotation of micro-organisms and of particles suspended in it. Its viscosity is twice as great at the freezing point as at ordinary summer temperature (77°F.).

Buoyancy—Water is a denser medium than air; it is 775 times heavier. Hence the buoyancy with which it supports a body immersed in it is correspondingly greater. The density of water is so nearly equal to that of protoplasm, that all living bodies will float in it with the aid of very gentle currents or of a very little exertion in swimming. Flying is a feat that only a few of the most specialized groups of animals have mastered, but swimming is common to all the groups.

Pressure—This greater density, however, involves greater pressure. The pressure is directly proportional to the depth, and is equal to the weight of the super-posed column of water. Hence, with increasing depth the pressure soon becomes enormous, and wholly insup-portable by bodies such as our own. Sponge fishers and pearl divers, thoroughly accustomed to diving, descending naked from a boat are able to work at depths up to 20 meters. Professional divers, encased in a modern diving dress are able to work at depths several times as great; but such depths, when compared with the depths of the great lakes and the oceans are com-parative shoals.

Beyond these depths, however, even in the bottom of the seas, animals live, adjusted to the great pressure, which may be that of several hundred of atmospheres. But these cannot endure the lower pressure of the surface, and when brought suddenly to the surface they burst. Fishes brought up from the bottom of the deeper freshwater lakes, reach the surface greatly swollen, their scales standing out from the body, their eyes bulging.

Maximum density—Water contracts on cooling, as do other substances, but not to the freezing point—only to 4° centigrade (39.2° Fahrenheit). On this peculiarity hang many important biological consequences. Below 4° C. it begins to expand again, becoming lighter, as shown in the accompanying table:

Temperature C°	F°	Weight in lbs. per cu. ft.	Density
35	95	62.060	.99418
21	70	62.303	.99802
10	50	62.408	.99975
4	39	62.425	1.00000
0	32	62.417	.99987

Hence, on the approach of freezing, the colder lighter water accumulates at the surface, and the water at the point of maximum density settles to the bottom, and the congealing process, so fatal to living tissues generally is resticted to a thin top layer. Here at 0° C. (32° F.) the water freezes, expanding about one-twelfth in bulk in the resulting ice and reducing its weight per cubic foot to 57.5 pounds.

Stratification of the water—Water is a poor conductor of heat. We recognize this when we apply heat to the bottom of a vessel, and set up currents for its distribution through the vessel. We depend on convection and not on conduction. But natural bodies of water are heated and cooled from the top, when they are in contact with the atmosphere and where the sun's rays strike. Hence, it is only those changes of temperature which increase the density of the surface waters that can produce convection currents, causing them to descend, and deeper waters to rise in their place. Minor changes of this character, very noticeable in shallow water, occur every clear day with the going down of the sun, but great changes, important enough to affect the temperature of all the waters of a deep lake, occur but twice a year, and they follow the precession of the equinoxes. There is a brief, often interrupted, period (in March in the latitude of Ithaca) after the ice has gone out, while the surface waters are being warmed to 0°C.; and there is a longer period in autumn, while they are being cooled to 0°C. Between times, the deeper waters of

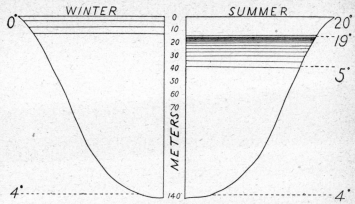

Fig. 4. Diagram illustrating summer and winter temperature conditions in Cayuga Lake. The spacing of the horizontal lines represents equal temperature intervals.

a lake are at rest, and they are regularly stratified according to their density.

In deep freshwater lakes the bottom temperature remains through the year constantly near the point of maximum density, 4° C. This is due to gravity. The heavier water settles, the lighter, rises to the top. Were gravity alone involved the gradations of temperature from bottom to top would doubtless be perfectly regular and uniform at like depths from shore to shore. But springs of ground water and currents come in to

Lake Temperatures 33

disturb the horizontal uniformity, and winds may do much to disturb the regularity of gradations toward the surface. Water temperatures are primarily dependent on those of the superincumbent air. The accompanying diagram of comparative yearly air and water temperatures in Hallstätter Lake (Austria) shows graphically the diminishing influence of the former on the latter with increasing depth.

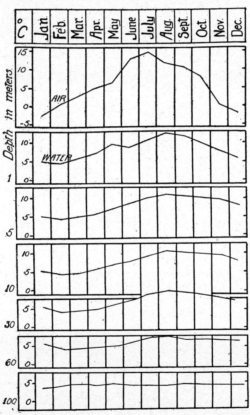

FIG. 5. Diagram illustrating the relation of air and water temperatures at varying depths of water in Hallstätter Lake (after Lorenz).

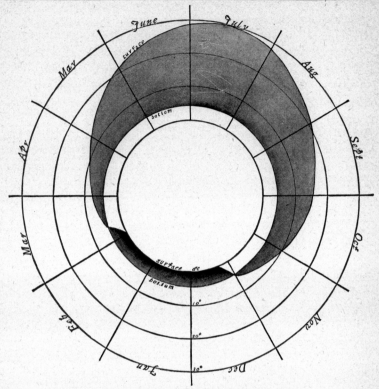

FIG. 6.   Diagram illustrating the distribution of temperature in Cayuga Lake
throughout the year.   (Extremes: not normal).

The yearly cycle—The general relation between surface and bottom temperatures for the year are graphically shown in the accompanying diagram, wherein the
two periods of thermal stratification, *"direct"* in summer
when the warmer waters are uppermost, and *"inverse"*
in winter when the colder waters are uppermost, are
separated by two periods of complete circulation, when
all the waters of the lake are mixed at 4° C.   The range
of temperatures from top to bottom is much greater in
the summer "stagnation period"; nevertheless there

is more real stagnation during the winter period; for, after the formation of a protecting layer of ice, this shuts out the disturbing influence of wind and sun and all the waters are at rest. The surface temperature bears no further relation to air temperature but remains constantly at 0° C.

After the melting of the ice in late winter the surface waters begin to grow warmer; so, they grow heavier, and tend to mingle with the underlying waters. When all the water in the lake is approaching maximum density strong winds heaping the waters upon a lee shore, may put the entire body of the lake into complete circulation. How long this circulation lasts will depend on the weather. It will continue (with fluctuating vigor) until the waters are warm enough so that their thermal stratification and consequent *resistance to mixture* are great enough to overcome the disturbing influence of the wind. Thereafter, the surface may be stirred by storms at any time, but the deeper waters of the lake will have passed into their summer rest.

On the approach of autumn the cooling of surface waters starts convection currents, which mix at first the upper waters only, but which stir ever more deeply as the temperature descends. When nearly 4°C., with the aid of winds, the entire mass of water is again put in circulation. The temperature is made uniform throughout, and what is more important biologically, the contents of the lake, in both dissolved and suspended matters, are thoroughly mixed. Nothing is thereafter needed other than a little further cooling of the surface waters to bring about the inverse stratification of the winter period.

Vernal and autumnal circulation periods differ in this, that convection currents have a smaller share, and winds may have a larger share in the former. For the surface waters are quickly warmed from 0° C. to 4° C., and further warming induces no descending currents, but instead tends toward greater stability. It sometimes happens that in shallow lakes there is little vernal circulation. If the water be warmed at 4° C. at the bottom before the ice is entirely gone, and if a period of calm immediately follow, so that no mixing is done by the wind, there may be no general spring circulation whatever.

The shallower the lake, other things being equal, the greater will be the departure of temperature conditions from those just sketched, for the greater will be the disturbing influence of the wind. In south temperate lakes, temperature conditions are, of course, reversed with the seasons. In tropical lakes whose surface temperature remains always above 4° C., there can be no complete circulation from thermal causes, and inverse stratification is impossible. In polar lakes, never freed from ice, no direct stratification is possible.

It follows from the foregoing that gravity alone may do something toward the warming of the waters in the spring, and much toward the cooling of them in the fall. By gravity they will be made to circulate until they reach the point of maximum density, when going either up or down the scale. Beyond this point, however, gravity tends to stabilize them. The wind is responsible for the further warming of the waters in early summer, and the heat in excess of 4° C. has been called by Birge and Juday "wind-distributed" heat. They estimate that it may amount to 30,000 gram-calories per square centimeter of surface in such lakes as those of Central New York, and the following figures for Cayuga Lake show its distribution by depth in August, 1911, in percentage remaining at successive ten-meter intervals below the surface:

Below	0	10	20	30	40	50	60	70	80	100	133 meters
%	100	50.2	16.7	7.1	3.7	2.4	1.8	1.2	.7	.3	remaining

These figures indicate the resistance to mixing that gravity imposes, and show that the wind is not able to overcome it below rather slight depths.

Vernal and autumnal periods of circulation have a very great influence upon the distribution of both organisms and their food materials in a lake; to the consideration of this we will have occasion to return later.

The thermocline—In the study of lake temperatures at all depths, a curious and interesting peculiarity of temperature interval has been commonly found pertaining to the period of direct stratification (midsummer). The descent in temperature is not regular from surface to bottom, but undergoes a sudden acceleration during a space of a very few meters some distance below the surface. The stratum of water in which this sudden drop of temperature occurs is known as the *thermocline* (German, *Sprungschicht*). It appears to represent the lower limit of the intermittent summer circulation due to winds. Above it the waters are more or less constantly stirred, below it they lie still. This interval is indicated by the shading on the right side of figure 4. Birge has designated the area above the thermocline as the epilimnion; the one below it as hypolimnion.

Further study of the thermocline has shown that it is not constant in position. It rises nearer to the surface at the height of the midsummer season and descends a few meters with the progress of the cooling of the autumnal atmosphere. This may be seen in figure 7, which is Birge and Juday's chart of temperatures of Lake Mendota as followed by them through the season of direct stratification and into the autumnal circulation period in 1906. This chart shows most graphically the growing divergence of surface and bottom temperatures up to August, and their later approximation and

final coalescence in October. Leaving aside the not
unusual erratic features of surface temperature (represented by the topmost contour line) it will be noticed
that there is a wider interval somewhere between 8 and
16 meters than any other interval either above or below
it. Sometimes it falls across two spaces and is rendered
less apparent in the charting by the selection of intervals. It first appears clearly in June at the 10–12 meter
interval. It rises in July above the 10 meter level.

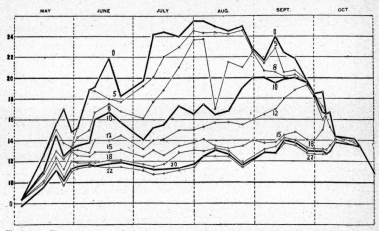

FIG. 7. Temperature of the water at different depths in Lake Mendota in
1906. The vertical spaces represent degrees Centigrade and the figures
attached to the curves indicate the depths in meters. (Birge and Juday).

In the middle of August it lies above the 8 meter level,
though it begins to descend later in the month. It
continues to descend through September, and is found
in early October between 16 and 18 meters. It disappears with the beginning of the autumnal circulation.
The cause of this phenomenon is not known. Richter
has suggested that convection currents caused by the
nocturnal cooling of the surface water after hot summer
days may be the cause of it. If the surface waters were

cooled some degrees they would descend, displacing the layers underneath and setting up shallow currents which would tend to equalize the temperature of all the strata involved therein. And if the gradation of temperatures downward were regular before this mixing, the result of it would be a sudden descent at its lower limit, after the mixing was done. This would account for the upper boundary of the thermocline, but not for its lower one. Perhaps an occasional deeper mixing, extending to its lower boundary, and due possibly to high winds, might bring together successional lower levels of temperature of considerable intervals. Perhaps the thermocline is but an accumulation of such sort of thermal disturbance-records, ranged across the vertical section of the lake, somewhat as wave-drift is ranged in a shifting zone along the middle of a sloping beach. At any rate, it appears certain that the thermocline marks the lower limit of the chief disturbing influences that act upon the surface of the lake. That it should rise with the progress of summer is probably due to the increasing stability of the lower waters, as differences in temperature (and therefore in density) between upper and lower strata are increased. Resistance to mixing increases until the maximum temperature is reached, · and thereafter declines, as the influence of cooling and of winds penetrates deeper and deeper.

In running water the mixing is more largely mechanical, and vertical circulation due to varying densities is less apparent. Yet the deeper parts of quiet streams approximate closely to conditions found in shallow lakes. Such thermal stratification as the current permits is direct in summer and inverse in winter, and there are the same intervening periods of thermal overturn when the common temperature approaches 4° C. In summer and in winter there is less "stagnation" of bottom waters owing to the current of the stream.

The thermal conservatism of water—Water is slower to respond to changes of temperature than is any other known substance. Its specific heat is greater. The heat it consumes in thawing (and liberates in freezing) is greater. The amount of heat necessary to melt one part of ice at 0° C. without raising its temperature at all would be sufficient to raise the temperature of the same when melted more than 75 degrees. Furthermore, the heat consumed in vaporization is still greater. The amount required to vaporize one part of water at 100° C. without raising its temperature would suffice to raise 534 parts of water from 0° C. to 1° C.; and the amount is still greater when vaporization occurs at a lower temperature. Hence, the cooling effect of evaporation on the surrounding atmosphere, which gives up its heat to effect this change of state in the water; hence, the equalizing effect upon climate of the presence of large bodies of water; hence the extreme variance between day and night temperatures in desert lands; hence the delaying of winter so long after the autumnal, and of summer so long after the vernal equinox. Water is the great stabilizer of temperature.

The content of natural waters—Water is the common solvent of all foodstuffs. These stuffs are, as everybody knows, such simple mineral salts as are readily leached out of the soil, and such gases as may be washed down out of the atmosphere. And since green plants are the producing class among organisms, all others being dependent on their constructive activities, water is fitted to be the home of life in proportion as it contains the essentials of green plant foods, with fit conditions of warmth, air and light.

Natural waters all contain more or less of the elementary foodstuffs necessary for life. Pure water (H_2O) is not found. All natural waters are mineralized waters—even rain, as it falls, is such. And a comparatively few soluble solids and gases furnish the still smaller number of chemical elements that go to make up the living substance. The amount of dissolved solids varies greatly, being least in rainwater, and greatest in dead seas, which, lacking outlet, accumulate salts through continual evaporation. Here is a rough statement of the dissolved solids in some typical waters:

In rain water	30—	40 parts per million			
In drainage water off siliceous soils	50—	80	"	"	"
In springs flowing from siliceous soils	60—	250	"	"	"
In drainage water off calcareous soils	140—	230	"	"	"
In springs flowing from calcareous soils	300—	660	"	"	"
In rivers at large	120—	350	"	"	"
In the ocean	33000—	37370	"	"	"

Thus the content is seen to vary with the nature of the soils drained, calcareous holding a larger portion of soluble solids than siliceous soils. It varies with presence or absence of solvents. Drainage waters from cultivated lands often contain more lime salts than do springs flowing from calcareous soils that are deficient in carbon dioxide. Spring waters are more highly charged than other drainage waters, because of prolonged contact as ground water with the deeper soil strata. And evaporation concentrates more or less the content of all impounded waters.

All natural waters contain suspended solids in great variety. These are least in amount in the well filtered water of springs, and greatest in the water of turbulent streams, flowing through fine soils. At the confluence of the muddy Missouri and the clearer Mississippi rivers the waters of the two great currents may be seen flowing together but uncommingled for miles.

The suspended solids are both organic and inorganic, and the organic are both living and dead, the latter being plant and animal remains. From all these non-living substances the water tends to free itself: The lighter organic substances (that are not decomposed and redissolved) are cast on shore; the heavier mineral substances settle to the bottom. The rate of settling is dependent on the rate of movement of the water and on the specific gravity and size of the particles. Fall Creek at Ithaca gives a graphic illustration of the carrying power of the current. In the last mile of its course, included between the Cornell University Campus and Cayuga Lake, it slows down gradually from a sheer descent of 78 ft. at the beautiful Ithaca Fall to a scarcely perceptible current at the mouth. It carries huge blocks of stone over the fall and drops them at its foot. It strews lesser blocks of stone along its bed for a quarter of a mile to a point where the surface ceases to break in riffles at low water. There it deposits gravel, and farther along, beds and bars of sand, some of which shift position with each flood rise, and consequent acceleration. It spreads broad sheets of silt about its mouth and its residual burden of finer silt and clay it carries out into the lake. The lake acts as a settling basin. Flood waters that flow in turbid, pass out clear.

Whipple has given the following figures for rate of settling as determined by size, specific gravity and form being constant:

Velocity of particles falling through water

Diameter 1.   inch, falls 100.   feet per minute.
   "   .1   "   "   8.   "   "   "
   "   .01   "   "   .15   "   "   "
   "   .001   "   "   .0015   "   "   "
   "   .0001   "   "   .000015   "   "   "

Suspended mineral matters are, as a rule, highly insoluble. Instead of promoting, they lessen the productivity of the water by shutting out the light.

Suspended organic solids likewise contribute nothing to the food supply as long as they remain undissolved. But when they decay their substance is restored to circulation. Only the dissolved substances that are in the water are at once available for food. The soil and the atmosphere are the great storehouses of these materials, and the sources from which they were all originally derived.

Gases from the atmosphere—The important gases derived from the atmosphere are two: carbon dioxide (CO_2) and oxygen (O). Nitrogen is present in the atmosphere in great excess (N, 79% to O, nearly 21%, and CO_2, .03%), and nitrogen is the most important constituent of living substance, but in gaseous form, free or dissolved, it is not available for food. The capacity of water for absorbing these gases varies with the temperature and the pressure, diminishing as warmth increases (insomuch that by boiling they are removed from it), and increasing directly as the pressure increases. Pure water at a pressure of 760 mm. in an atmosphere of pure gas, absorbs these three as follows:

	Oxygen	CO₂	Nitrogen
At 0°C	41.14	1796.7	20.35
At 20°C	28.38	901.4	14.03

At double the pressure twice the quantity of the gas would be dissolved. Natural waters are exposed not to the pure gas but to the mixture of gases which make up the atmosphere. In such a mixture the gases are absorbed independently of each other, and in proportion to their several pressures, which vary as their several densities: the following table* shows, for

---

*Abridged from a table of values to tenths of a degree by Birge and Juday in Bull. 22, Wisc. Geol. & Nat. Hist. Survey, p. 20.

example, the absorbing power of pure water at various temperatures for oxygen from the normal atmosphere at 760 mm. pressure:

Water at 0°C 9.70 cc. per liter   at 15°C 6.96 cc. per liter
   "   5°C 8.68 cc. "   "   " 20°C 6.28 cc. "   "
   "   10°C 7.77 cc. "   "   " 25°C 5.76 cc. "   "

The primary carbon supply for the whole organic world is the carbon dioxide (CO_2) of the atmosphere. Chlorophyll-bearing plants are the gatherers of it. They alone among the organisms are able to utilize the energy of the sun's rays. The water existing as vapor in the atmosphere is the chief agency for bringing these gases down to earth for use. Standing water absorbs them at its surface but slowly. Water vapor owing to better exposure, absorbs them to full saturation, and then descends as rain. In fresh water they are found in less varying proportion, varying from none at all to considerable degree of supersaturation. Birge and Juday report a maximum occurrence of oxygen as observed in the lakes of Wisconsin of 25.5 cc. per liter in Knight's Lake on Aug. 26, 1909 at a depth of 4.5 meters. This water when brought to the surface (with consequent lowering of pressure by about half an atmosphere) burst into lively effervescence, with the escape of a considerable part of the excess oxygen into the air. ('11, p. 52). They report the midsummer occurrence of free carbon dioxide in the bottom waters of several lakes in amounts approaching 15 cc. per liter.

The reciprocal relations of CO_2 and O—Carbon dioxide and oxygen play leading roles in organic metabolism, albeit, antithetic roles. The process begins with the cleavage of the carbon dioxide, and the building up of its carbon into organic compounds; it ends with the oxidation of effete carbonaceous stuffs and the reappearance of CO_2. Both are used over and over again.

Plants require CO_2 and animals require oxygen in order to live and both live through the continual exchange of these staple commodities. This is the best known phase in the cycle of food materials. The oxygen is freed at the beginning of the synthesis of organic matter, only to be recombined with the carbon at the end of its dissolution. And the well-being of the teeming population of inland waters is more dependent on the free circulation and ready exchange of the dissolved supply of these two gases than on the getting of a new supply from the air.

The stock of these gases held by the atmosphere is inexhaustible, but that contained in the water often runs low; for diffusion from the air is slow, while consumption is sometimes very rapid. We often have visible evidence of this. In the globe in our window holding a water plant, we can see when the sun shines streams of minute bubbles of oxygen, arising from the green leaves. Or, in a pond we can see great masses of algae floated to the surface on a foam of oxygen bubbles. We cannot see the disappearance of the carbon dioxide but if we test the water we find its acidity diminishing as the carbon dioxide is consumed.

At times when there is abundant growth of algae near the surface of a lake there occurs a most instructive diurnal ebb and flow in the production of these two gases. By day the well lighted layers of the water become depleted of their supply of CO_2 through the photosynthetic activities of the algae, and become supersaturated with the liberated oxygen. By night the microscopic crustaceans and other plancton animals rise from the lower darker strata to disport themselves nearer the surface. These consume the oxygen and restore to the water an abundance of carbon dioxide. And thus when conditions are right and the numbers of

plants and animals properly balanced there occur
regular diurnal fluctuations corresponding to their
respective periods of activity in these upper strata.

Photosynthesis is, however, restricted to the better
lighted upper strata of the water. The region of
greatest carbon consumption is from one to three meters
in depth in turbid waters, and of ten meters or more in
depth in clear lakes. Consumption of oxygen, however,
goes on at all depths, wherever animal respiration or
organic decomposition occurs. And decomposition
occurs most extensively at the bottom where the organic
remains tend to be accumulated by gravity. With a
complete circulation of the water these two gases may
continue to be used over and over again, as in the example just cited. But, as we have seen, there is no circulation of the deeper water during two considerable periods
of the year; and during these stagnation periods the
distribution of these gases in depth becomes correlated
in a wonderful way with the thermal stratification of
the water. This has been best illustrated by the work
of Birge and Juday in Wisconsin. Figure 8 is their
diagram illustrating the distribution of free oxygen in
Mendota Lake during the summer of 1906. It should
be studied in connection with figure 7, which illustrates
conditions of temperature. Then it will be seen that
the two periods of equal supply at all levels correspond
to vernal and autumnal circulation periods. The
season opens with the water nearly saturated (8 cc. of
oxygen per liter of water) throughout. With the warming of the waters the supply begins to decline, being
consumed in respiration and in decomposition. In the
upper six or seven meters the decline is not very extensive, for at these depths the algae continually renew the
supply. But as the lower strata settle into their summer rest their oxygen content steadily disappears, and
is not renewed until the autumnal overturn. For three

months there is no free oxygen at the bottom of the lake, and during August there is not enough oxygen below the ten meter level to keep a fish alive.

Correspondingly, the amount of free CO_2 in the deeper strata of the lake increases rather steadily until the autumnal overturn. It is removed from circulation, and in so far as it is out of the reach of effective light, it is unavailable for plant food.

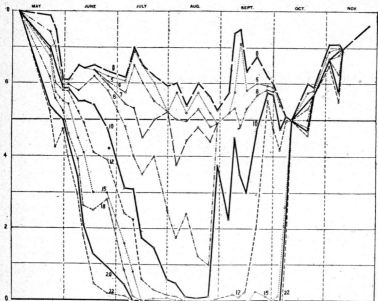

FIG. 8. Dissolved oxygen at different depths in Lake Mendota in 1906. The vertical spaces represent cubic centimeters of gas per liter of water and the figures attached to the curves indicate the depths in meters. (Birge and Juday.)

Other gases—A number of other gases are more or less constantly present in the water; nitrogen, as above stated, being absorbed from the air, methane (CH_4), and other hydrocarbons, and hydrogen sulphide (H_2S), etc., being formed in certain processes of decom-

position. Of these, methane or marsh gas, is perhaps
the most important. This is formed where organic
matter decays in absence of oxygen. In lakes such
conditions are found mainly on the bottom. In marshes
and stagnant shoal waters generally, where there is
much accumulation of organic matter on the bottom,
this gas is formed in abundance. It bubbles up through
the bottom ooze, or often buoys up rafts of agglutinated
bottom sediment.

Nitrogen—The supply of nitrogen for aquatic organisms is derived from soluble simple nitrates (KNO_3,
$NaNO_3$, etc.) Green plants feed on these, and build
proteins out of them. And when the plants die (or
when animals have eaten them) their dissolution yields
two sorts of products, ammonia and nitrates, that
become again available for plant food. Ammonia is
produced early in the process of decay and the nitrates
are its end products.

Bacteria play a large role in the decomposition of
proteins. At least four groups of bacteria successively
participate in their reduction. The first of these are
concerned with the liquefaction of the proteins, hydroly-
zing the albumins, etc., by successive stages to albu-
moses, peptones, etc., and finally to ammonia. A
second group of bacteria oxidizes the ammonia to
nitrites. A third group oxidizes the nitrites to
nitrates. A fourth group, common in drainage waters,
reduces nitrates to nitrites. Since these processes are
going on side by side, nitrogen is to be found in all
these states of combination when any natural water is
subjected to chemical analysis. The following table
shows some of the results of a large number (415) of
analyses of four typical bottomland bodies of water,
made for Kofoid's investigation of the plancton of the
Illinois River by Professor Palmer.

The relative productiveness in open-water life of these situations is shown in the last column of the table.

In parts per million	Solids		Free Ammonia	Organic Nitrogen	Nitrites	Nitrates	Plancton cm3 per m3
	Suspended	Dissolved					
Illinois River .	61.4	304.1	.860	1.03	.147	1.59	1.91
Spoon River ..	274.3	167.1	.245	1.29	.039	1.01	.39
Quiver Lake .	25.1	248.2	.165	.61	.023	.66	1.62
Thompson's L.	44.6	282.9	.422	1.05	.048	.64	6.68

The difference between these four adjacent bodies of water explains some of the peculiarities of the table. The rivers hold more solids in suspension than do the lakes, although these lakes are little more than basins holding impounded river waters. Spoon River holds the least amount of dissolved solids, and by far the greatest amount of suspended solids. Since the latter are not available for plant food, naturally this stream is least productive of plancton. Illinois River drains a vast and fertile region, and receives in its course the sewage and other organic wastes of two large cities, Chicago and Peoria, and of many smaller ones. Hence, its high content of dissolved matter, the cities being remote, so there has been time for extensive liquefaction. Hence, also, its high content of ammonia, of nitrites and of nitrates.

The two lakes are very unlike; Quiver Lake is a mere strip of shoal water, fed by a clear stream that flows in through low sandy hills. It receives water from the Illinois River only during high floods. Thompson's Lake is a much larger body of water, fed directly from the Illinois River through an open channel. Naturally, it is much like the river in its dissolved solids, and in its total organic nitrogen. That it falls far below the river in nitrates and rises high above it in plancton production may perhaps be due to the extensive con-

sumption of nitrates by plancton algae.   Nitrates, because they furnish nitrogen supply in the form at once
available for plant growths, are, in shallow waters at
least, an index of the fertility of the water.   As on
land, so in the water, the supply of these may be
inadequate for maximum productiveness, and they may
be added with profit as fertilizer.

The carbonates—Lime and magnesia combine with
carbon dioxide, abstracting it from the water, forming

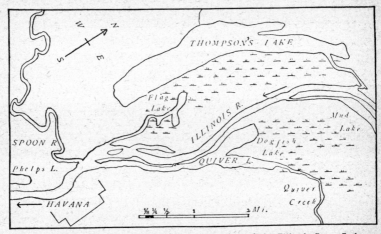

FIG. 9.   Environs of the Biological Field Station of the Illinois State Laboratory of Natural History, the scene of important work by Kofoid and others
on the life of a great river.

solid carbonates ($CaCO_3$ and $MgCO_3$).   These accumulate in quantities in the shells of molluscs, in the stems
of stoneworts, in the incrustations of certain pond
weeds, and of lime-secreting algae.   The remains of
such organisms accumulate as marl upon the bottom.
The carbonates (and other insoluble minerals) remain;
the other body compounds decay and are removed.
By such means in past geologic ages the materials for
the earth's vast deposits of limestone were accumu-

lated. Calcareous soils contain considerable quantities of these carbonates.

In pure water these simple carbonates are practically insoluble; but when carbon dioxide is added to the water, they are transformed into bicarbonates* and are readily dissolved.† So the carbonates are leached out of the soils and brought back into the water. So the solid limestone may be silently removed, or hollowed out in great caverns by little underground streams. So the Mammoth Cave in Kentucky, and others in Cuba, in Missouri, in Indiana and elsewhere on the continent, have been formed.

The water gathers up its carbon dioxide in part as it descends through the atmosphere, and in larger part as it percolates thru soil where decomposition is going on and where oxidation products are added to it.

Carbon dioxide, thus exists in the water in three conditions: (1) Fixed (and unavailable as plant food) in the simple carbonates; (2) "half-bound" in the bicarbonates; and (3) free. Water plants use first for food, the free carbon dioxid, and then the "half bound" that is in loose combination in the bicarbonates. As this is used up the simple carbonates are released, and the water becomes alkaline.§ Birge and Juday have several times found a great growth of the desmid *Staurastrum* associated with alkalinity due to this cause. In a maximum growth which occurred in alkaline waters at a depth of three meters in Devil's Lake, Wisconsin, on June 15th, 1907, these plants numbered 176,000 per liter of water.

---

*$CaCO_3$, for example, becoming $Ca(HCO_3)_2$, the added part of the formula representing a molecule each of CO_2 and H_2O.

†If "hard" water whose hardness is due to the presence of these bicarbonates be boiled, the CO_2 is driven off and the simple carbonates are re-precipitated (as, for example, on the sides and bottom of a tea kettle). This is "temporary hardness." "Permanent hardness" is due to the presence of sulphates and chlorides of lime and magnesia, which continue in solution after boiling.

§Phenolphthalein, being used as indicator of alkalinity.

Waters that are rich in calcium salts, especially in calcium carbonate, maintain, as a rule, a more abundant life than do other waters. Especially favorable are they to the growth of those organisms which use much lime for the building of their hard parts, as molluscs, stoneworts, etc. There are, however, individual preferences in many of the larger groups. The crustaceans for example, prefer, as a rule, calcium rich waters, but one of them, the curious entomostracan, *Holopedium gibberum*, (Fig. 10) is usually found in calcium poor waters, in lakes in the Rocky Mountains and in the Adirondacks, in waters that flow off archæan rocks or out of siliceous sands. The desmids with few exceptions are more abundant in calcium poor waters. The elegant genus Micrasterias is at Ithaca especially abundant in the peat-stained calcium-poor waters of sphagnum bogs.

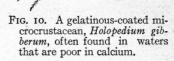

FIG. 10. A gelatinous-coated microcrustacean, *Holopedium gibberum*, often found in waters that are poor in calcium.

Other minerals in the water—The small quantities of other mineral substances required for plant growth are furnished mainly by a few sulphates, phosphates and chlorids: sulphates of sodium, potassium, calcium and magnesium; phosphates of iron, aluminum, calcium and magnesium, and chlorids of sodium, potassium, calcium and magnesium. Aluminum alone of the elements composing the above named compounds, is not always requisite for growth, although it is very often present. Silica, likewise, is of wide distribution, and occurs in the water in considerable amounts, and is used by many organisms in the growth of their hard parts. As the stoneworts use lime for their growth, some 4% of the dry weight of Chara being CaO, so diatoms require silica to build their shells. When the diatoms are dead their shells, relatively heavy though extremely minute, slowly settle to the bottom, slowly dissolving; and so, analyses of lake waters taken at different depths usually show increase of silica toward the bottom.

Iron, common salt, sulphur, etc., often occur locally in great abundance, notably in springs flowing from special deposits, and when they occur they possess a fauna and flora of marked peculiarities and very limited extent.

An idea of the relative abundance of the commoner mineral substances in lake waters may be had from the following figures that are condensed from Birge and Juday's report of 74 analyses.

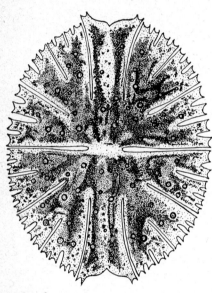

FIG. 11. A beautiful green desmid, *Micrasterias* that is common in bog waters.

MINERAL CONTENT OF WISCONSIN LAKES

Parts per million

	SiO_2	Fl_2O_3 + Al_3O_3	Ca	Mg	Na	K	CO_3	HCO_3	SO_4	Cl
Minimum	0.8	0.4	0.6	0.3	0.3	0.3	0.0	4.9	0.0	1.5
Maximum	33.0	11.2	49.6	32.7	6.2	3.1	12.0	153.0	18.7	10.0
Average	11.7	2.1	26.9	19.6	3.2	2.2	2.1	91.7	9.8	3.9

This is the bill of fare from which green water plants may choose. Forel aptly compared the waters of a lake to the blood of the animal body. As the cells of the body take from the blood such of its content as is suited to their need, so the plants and animals of the water renew their substance out of the dissolved substances the water brings to them.

Organic substances dissolved in the water may so affect both its density and its viscosity as to determine both stratification and distribution of suspended solids. This is a matter that has scarcely been noticed by limnologists hitherto. Dr. J. U. Lloyd ('82) long ago showed how by the addition of colloidal substances to a vessel of water the whole contents of the vessel can be broken into strata and these made to circulate, each at its own level, independent of the other strata. Solids in suspension can be made to float at the top of particular strata, according to density and surface tension.

Perhaps the "false bottom" observed in some northern bog-bordered lakes is due to the dissolved colloids of the stratum on which it floats. Holt ('08, p. 219) describes the "false bottom" in Sumner Lake, Isle Royal, as lying six to ten feet below the surface, many feet above the true bottom; as being so tenuous that a pole could be thrust through it almost as readily as through clear water; and as being composed of fine disintegrated remains of leaves and other light organic material. "In places there were great breaks in the 'false bottom,' doubtless due to the escape of gases which had lifted this fine ooze-like material from a greater depth: and through these breaks one could look down several feet through the brownish colored water."

Perhaps the colloidal substances in solution are such as harden upon the surface of dried peat, like a water-proof glue, making it for a time afterward impervious to water.

WATER AND LAND

OCEANS are the earth's great storehouse of water. They cover some eight-elevenths of the surface of the earth to an average depth of about two miles. They receive the off-flow from all the continents and send it back by way of the atmosphere.

The fresh waters of the earth descend in the first instance out of the atmosphere. They rise in vapor from the whole surface of the earth, but chiefly from the ocean. Evaporation frees them from the ocean's salts, these being non-volatile. They drift about with the currents of the atmosphere, gathering its gases to saturation, together with very small quantities of drifting solids; they descend impartially upon water and land, chiefly as rain, snow and hail.

They are not distributed uniformly over the face of the continents for each continent has its humid regions and its deserts. Rainfall in the United States varies from 5 to 100 inches per annum. Two-thirds of it falls on the eastern three-fifths of the country. For the Eastern United States it averages about 48 inches, for the Western United States about 12 inches; the average for the whole is about 30 inches. The total annual precipitation is about 5,000,000,000 acre-feet.*

*An acre-foot is an acre of water 1 foot deep or 43,560 cubic feet of water.

It is commonly estimated that at least one-half of this rainfall is evaporated, in part from soil and water surfaces, but much more from growing vegetation; for the transpiration of plants gives back immense quantities of water to the atmosphere. Hellriegel long ago showed that a crop of corn requires 300 tons of water per acre: of potatoes or clover, 400 tons per acre. At the Iowa Agricultural Experiment Station it was shown that an acre of pasturage requires 3,223 tons of water, or 28 inches in depth ($2\frac{1}{3}$ acre-feet). Before the days of tile drainage it was a not uncommon practice to plant willow trees by the edges of swales, in order that they might carry off the water through their leaves, leaving the ground dry enough for summer cropping. The rate of evaporation is accelerated also, by high temperatures and strong winds.

The rain tends to wet the face of the ground everywhere. How long it will stay wet in any given place will depend on topography and on the character of the soil as well as on temperature and air currents. Showers descending intermittently leave intervals for complete run-off of water from the higher ground, with opportunity for the gases of the atmosphere to enter and do their work of corrosion. The dryer intervals, therefore, are times of preparation of the materials that will appear later in soil waters. Yet all soils in humid regions retain sufficient moisture to support a considerable algal flora. Periodical excesses of rainfall are necessary also to maintain the reserve of ground water in the soil. Suppose, for example, that the 35 inches of annual rainfall at Ithaca were uniformly distributed. There would be less than one-tenth of an inch of precipitation each day—an amount that would be quickly and entirely evaporated, and the ground would never be thoroughly wet and there would be no ground water to replenish the streams. Storm waters

tend to be gathered together in streams, and thus about
one-third of our rainfall runs away.   In humid areas
small streams converge to form larger ones, and flow
onward to the seas.   In arid regions they tend to
spread out in sheet floods, and to disappear in the sands.

In a state of nature little rain water runs over the
surface of the ground, apart from streams.   It mainly
descends into the soil.   How much the soil can hold
depends upon its composition.   Dried soils have a
capacity for taking up and holding water about as follows: sharp sand 25%, loam 50%, clay 60%, garden
mould 90% and humus 180% of their dry weight.   Water
descends most rapidly through sand and stands longest
upon the surface of pure clay.   Thick vegetation with
abundant leaf fall, and humus in the soil tend to hinder
run-off of storm waters, and to prolong their passage
through the soil. Thus the excess of rainfall is gradually
fed into the streams by springs and seepage.   Under
natural conditions streams are usually clear, and their
flow is fairly uniform.

Unwise clearing of the land and negligent cultivation
of the soil facilitate the run-off of the water before the
storm is well spent, promote excessive erosion and
render the streams turbid and their volume abnormally
fluctuating.   Little water enters the soil and hence the
springs dry up, and the brooks, also, as the seepage of
ground water ceases.   Two great evils immediately
befall the creatures that live in the streams and pools:
(1) There is wholesale direct extermination of them with
the restriction of their habitat at low water.   (2) There
occurs smothering of them under deposits of sediment
brought down in time of floods, with indirect injury to
organisms not smothered, due to the damage to their
foraging grounds.

The waters of normal streams are derived mainly
from seepage, maintained by the store of water accumu-

lated in the soil. This store of ground water amounts according to recent estimates to some 25% of the bulk of the first one hundred feet in soil depth. Thus it equals a reservoir of water some 25 feet deep covering the whole humid eastern United States. It is continuous over the entire country. Its fluctuations are studied by means of measurements of wells, especially by recording the depth of the so-called "water table." On the maintenance of ground water stream-flow and organic productiveness of the fields alike depend.

CHAPTER III
TYPES OF AQUATIC
ENVIRONMENT

I. LAKES AND
PONDS

OUT of the atmosphere comes our water supply —the greatest of our natural resources. It falls on hill and dale, and mostly descends into the soil. The excess off-flowing from the surface and outflowing from springs and seepage, forms water masses of various sorts according to the topography of the land surface. It forms lakes, streams or marshes according as there occur basins, channels or only plant accumulations influencing drainage.

The largest of the bodies of water thus formed are the lakes. Our continent is richly supplied with them, but they are of very unequal distribution. The lake regions in America as elsewhere are regions of comparatively recent geological disturbance. Lakes thickly dot the peninsula of Florida, the part of our continent most recently lifted from the sea. Over the northern recently glaciated part of the continent they are innumerable, but in the great belts of corn and cotton, and on the plains to the westward, they are few and far between. They are abundant in the regions of more recent volcanic disturbance in our western mountains, but are practically absent from the geologically older Appalachian hills. They lie in the depressions between the recently uplifted lava blocks of southern Oregon. They occur also in the craters of extinct volcanoes. They are apt to be most picturesque when their setting is in the midst of mountains. There are probably no more beautiful lakes in the world than some of those in the West, such as Lake Tahoe (altitude 6200 ft.) on the California-Nevada boundary, and Lake Chelan in the state of Washington*, to say nothing of the Coeur d'Alene in Idaho and Lake Louise in British Columbia. Eastward the famous lake regions that attract most visitors are those of the mountains of New York and New England, those of the woodlands of Michigan and Wisconsin and those of the vast areas of rocks and water in Canada.

Lakes are temporary phenomena from the geologists point of view. No sooner are their basins formed than the work of their destruction begins. Water is the agent of it, gravity the force employed, and erosion the chief method. Consequently, other things being equal, the processes of destruction go on most rapidly in regions of abundant rainfall. Inwash of silt from surrounding slopes tends to fill up their basins. The most extensive filling is about the mouths of inflowing streams, where mud flats form, and extend in Deltas out into the lake. These deltas are the exposed summits of great mounds of silt that spread out broadly underneath the water on the lake floor. At the shorelines these deposits are loosened by the frosts of winter,

---

*Descriptions of these two lakes will be found in Russell's *Lakes of North America*.

pushed about by the ice floes of spring, and scattered by every summer storm, but after every shift they settle again at lower levels. Always they are advancing and filling the lake basin. The filling may seem slow and insignificant on the shore of one of the Great Lakes but its progress is obvious in a mill pond, and the difference is only relative.

Fig. 12. An eroding bluff on the shore of Lake Michigan that is receding at the rate of several feet each year. The broad shelving beach in the foreground is sand, where the waves ordinarily play. Against the bare rising boulder-strewn strip back of this, the waves beat in storms; at its summit they gather the earth-slides from the bank above and carry them out into the lake. The black strip at the rear of the sand is a line of insect drift, deposited at the close of a midsummer storm by the turning of the wind on shore.

On the other hand, lakes disappear with the cutting down of the rim of their basins in outflow channels. The Niagara river, for example, is cutting through the limestone barrier that retains Lake Erie. At Niagara Falls it is making progress at the rate of about five feet a year. Since the glacial period it has cut back from the shore of Lake Ontario a distance of some seventeen miles, and if the process continues it will in time empty Lake Erie.

Fig. 13. Evans' Lake, Michigan; a lake in process of being filled by encroachment of plants. A line of swamp loose-strife (*Decodon*) leads the invading shore vegetation. Further inwash of silt or lowering of outlet is precluded by density of the surrounding heath. The plants control its fate.

Photo by E. McDonald.

When the glacier lay across the St. Lawrence valley, before it had retreated to the northward, all the waters of the great lakes region found their way to the ocean through the Mohawk Valley and the Hudson. At that time a similar process of cutting an outlet through a limestone barrier was going on near the site of the present village of Jamesville, New York, where on the

Clark Reservation one may see today a series of abandoned cataracts, dry rock channels and plunge basins. Green Lake at present occupies one of these old plunge basins, its waters, perhaps a hundred feet deep, are surrounded on all sides but one, by sheer limestone cliffs nearly two hundred feet high.

When lakes become populated then the plants and animals living in the water and about the shore line contribute their remains to the final filling of the basin. This is well shown in figure 13.

The Great Lakes constitute the most magnificent system of reservoirs of fresh water in the world; five v a s t i n l a n d seas, whose shores have all the sweep and majesty of the ocean, no land being visible across them. All but one (Erie) have the bottom of their basins below the sea level. Their area, elevation and depth are as follows:

Fig. 14.   The larger lakes and rivers of North America.

	Area in sq. mi.	Surface alt. in ft.	Depth in feet mean†	maximum
Lake Ontario	7.240	247	300	738
" Erie	9.960	573	70	210
" Huron*	23.800	581	250	730
" Michigan	22.450	581	325	870
" Superior	31.200	602	475	1.008

*Including Georgian Bay.
†Approximate.

They are stated by Russell to contain enough water to keep a Niagara full-flowing for a hundred years.

The Finger Lakes of the Seneca basin in Central New York constitute an unique series occupying one section of the drainage area of Lake Ontario, with which they communicate by the Seneca and Oswego rivers. They occupy deep and narrow valleys in an upland plateau of soft Devonian shales. Their shores are rocky and increasingly precipitous near their southern ends. The marks of glaciation are over all of them. Keuka, the most picturesque of the series, occupies a forking valley partially surrounding a magnificent ice-worn hill. The others are all long and narrow and evenly contoured, without islands (save for a single rocky islet near the east Cayuga shore) or bays.

The basins of these lakes invade the high hills to the southward, reaching almost to the head-waters of the tributaries of the Susquehanna River. Here there is found a wonderful diversity of aquatic situation. At the head of Cayuga Lake, for example, beyond the deep water there is a mile of broad shelving silt-covered lake bottom, ending in a barrier reef. Then there is a broad flood plain, traversed by deep slow meandering streams, and covered in part by marshes. Then come the hills, intersected by narrow post-glacial gorges, down which dash clear streams in numerous beautiful waterfalls and rapids. Back of the first rise of the hills the streams descend more slowly, gliding along over pebbly beds in shining riffles, or loitering in leaf-strewn woodland pools. A few miles farther inland they find their sources in alder-bordered brooks flowing from sphagnum bogs and upland swales and springs.

Thus the waters that feed the Finger Lakes are all derived from sources that yield little aquatic life, and they run a short and rapid course among the hills, with little time for increase by breeding: hence they contribute little to the population of the lake. They bring in constantly, however, a supply of food materials, dissolved from the soils of the hills.

Bordering the Finger Lakes there are no extensive marshes, save at the ends of Cayuga, and the chief irregularities of outline are formed by the deltas of inflowing streams. The two large central lakes, Cayuga and Seneca, have their basins extending below the sea level. Their sides are bordered by two steeply-rising, smoothly eroded hills of uniform height, between which they lie extended like wide placid rivers. The areas, elevations and depths of the five are as follows:

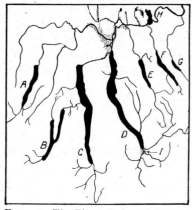

Fig. 15. The Finger Lakes of Central New York.

A, Canandaigua; B, Keuka; C, Seneca; D, Cayuga; E, Owasco; F, Skaneateles; G, Otisco; H, the Seneca River; I, The arrow indicates the location of the Cornell University Biological Field Station at Ithaca. The stippled area at the opposite end of Cayuga Lake marks the location of the Montezuma Marshes.

	Area sq. mi.	Surface alt. in ft.	Depth in feet mean	maximum
Lake Skaneateles	13.9	867	142	297
" Owasco	10.3	710	95	177
" Cayuga	66.4	381	177	435
" Seneca	67.7	444	288	618
" Keuka	18.1	709	99	183
" Canandaigua	16.3	686	126	274

Birge and Juday found the transparency of four of these lakes as measured by Secchi's disc in August, 1910, to be as follows:

Canandaigua ..... 12.0 ft.   Seneca .... 27.0 ft.
Cayuga .... 16.6 ft.   Skaneateles .... 33.5 ft.

The Lakes of the Yahara Valley in Southern Wisconsin are of another type. They occupy broad, shallow basins formed by the deposition of barriers of glacial drift in the preglacial course of the Yahara River. Their outlet is through Rock River into the Mississippi. Their shores are indented with numerous bays, and bordered extensively by marshes. The surrounding plain is dotted with low rounded hills, some of which rise abruptly from the water, making attractive shores. The city of Madison is the location of the University of Wisconsin, which Professor Birge has made the center of the most extensive and careful study of lakes yet undertaken in America. The area, elevation and depth of these lakes is as follows:

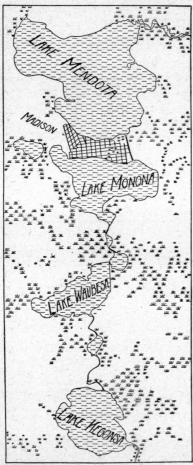

FIG. 16. The four-lake region of Madison, Wisconsin.

	Area in sq. mi.	Surface alt. in ft.	Depth in feet mean	maximum
Lake Kegonsa	15	842	15	31
" Wabesa	3	844	15	36
" Monona	6	845	27	75
" Mendota	15	849	40	85

Lakes resulting from Erosion—Although erosion tends generally to destroy lakes by eliminating their basins, here and there it tends to foster other lakes by making basins for them. Such lakes, however, are shallow and fluctuating. They are of two very different sorts, *floodplain lakes* and *solution lakes*.

Floodplain Lakes and Ponds—Basins are formed in the floodplains of rivers by the deposition of barriers of eroded silt, in three different ways.

1. By the deposition across the channel of some large stream of the detritus from a heavily silt-laden tributary stream. This blocks the larger stream as with a partial dam, creating a lake that is obviously but a dilatation of the larger stream. Such is Lake Pepin in the Mississippi River, created by the barrier that is deposited by the Chippewa River at its mouth.

2. By the partial filling up of the abandoned channels of rivers where they meander through broad alluvial bottom-lands. Phelps Lake partly shown in the figure on page 50 is an example of a lake so formed; and all the other lakes of that figure are partly occluded by similar deposits of river silt. Horseshoe bends are common in slow streams, and frequently a river will cut across a bend, shortening its course and opening a new channel; the filling up with silt of the ends of the abandoned channel results in the formation of an "ox-bow" lake; such lakes are common along the lower course of the Mississippi, as one may see by consulting any good atlas.

3. By the deposition in times of high floods of the bulk of its load of detritus at the very end of its course, where it spreads out in the form of a delta. Thus a barrier is often formed on one or both sides, encircling a broad shallow basin. Such is Lake Pontchartrain at the left of the ever extending delta of the Mississippi.

Solution Lakes and Ponds—Of very different character are the lakes whose basins are produced by the dissolution of limestone strata and the descent of the overlying soil in the form of a "sink." This is erosion, not by mechanical means at first, but by solution. It

FIG. 17. Solution lakes of Leon County, Florida, (after Sellards).

The white spots in the lakes indicate sinks.
A. Lake Iamonia; area at high water 10 sq. mi.
B. Lake Jackson; area 7 sq. mi.
C. Lake Lafayette; area 3½ sq. mi.
D. Lake Miccosukee; area 7½ sq. mi.; depth of north sink 28 ft. Water escapes through this sink at the estimated rate of 1000 gals. per minute.
O. Ocklocknee River; S, St. Mark's River; T, Tallahassee.

occurs where beds of soluble strata lie above the permanent ground water level, and are themselves overlaid by clay. Rain water falling through the air gathers carbon dioxide and becomes a solvent of limestone. Percolating downward through the soil it passes through the permeable carbonate, dissolving it and carrying its substance in solution to lower levels, often flowing out in springs. As the limestone is thus removed the superincumbent soil falls in, forming a sink hole. The widening of the hole, by further solution and slides results in the formation of the pond or lake, possibly, at the beginning, as a mere pool.

The area of such a lake is doubtless gradually increased by the settling of the bottom around the sink as the soluble limestone below is slowly carried away. Its configuration is in part determined by the original topography of the land surface, and in part by the course of the streamflow underground: but its bed is unique among lake bottoms in that all its broad shoals suddenly terminate in one or more deep funnel-shaped outflow depressions.

Lime sinks occur over considerable areas in the southern states, and in those of the Ohio Valley, but perhaps the best development of lakes about them is in the upland region of northern Florida. These lakes are shallow basins having much of their borders ill-defined and swampy. Perhaps the most remarkable of them is Lake Alachua near Gainesville. At high water this lake has an area of some twenty-five square miles and a depth (outside the sink) of from two to fourteen feet. At its lowest known stage it is reduced to pools filling the sinks. During its recorded history it has several times alternated between these conditions. It has been for years a vast expanse of water carrying steamboat traffic, and it has been for other years a broad grassy plain, with no water in sight. The widening or the stoppage of the sinks combined with excessive or scanty rainfall have been the causes of these remarkable changes of level.

The sinks are more or less funnel-shaped openings

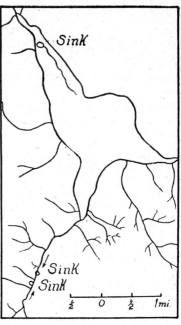

FIG. 18. Lake Miccosukee, (after Sellards), showing sinks; one in lake bottom at north end, two in outflowing stream, 2½ miles distant. Arrows indicate normal direction of stream flow, (reversed south of sinks in flood time when run-off is into St. Mark's River).

leading down through the soil into the limestone. Ditchlike channels often lead into them across the lake's bottom. The accompanying diagram shows that they are sometimes situated outside the lake's border, and suggest that such lakes may originate through the formation of sinks in the bed of a slow stream.

Such lakes, when their basins lie above the level of the permanent water table, may sometimes be drained by sinking wells through the soil of their beds. This allows the escape of their waters into the underlying limestone. Sometimes they drain themselves through the widening of their underground water channels. Always they are subject to great changes of level consequent upon variation in rainfall.

Enough examples have now been cited to show how great diversity there is among the fresh-water lakes of North America. Among those we have mentioned are the lakes that have received the most attention from limnologists hitherto; but hardly more than a beginning has been made in the study of any of them. Icthyologists have collected fishes from most of the lakes of the entire continent, and plancton collections have been made from a number of the more typical: from Yellowstone Lake by Professor Forbes in 1890 and from many other lakes, rivers and cave streams since that date.

Lakeside laboratories—On the lakes above mentioned are located a number of biological field stations. That at Cornell University is at the head of Cayuga Lake. That of the Ohio State University is on Gibraltar Island in Lake Erie, near Put-in-Bay, Ohio. That of the University of Pittsburgh is on the shore of the same lake at Erie, Pennsylvania. The biological laboratories of the University of Wisconsin are located directly upon the shore of Lake Mendota, and a special Lake Limnological Laboratory is maintained at Trout Lake. The University of Florida at Gainesville is conveniently near to a number of the solution lakes of northern Florida. Elsewhere there are other lakeside research stations among which we may mention the following:

That of the University of Michigan is on Douglas Lake in the northern end of the southern peninsula of Michigan.

That of the University of Indiana is on Winona Lake, a shallow hard water lake of irregular outline, having an area of something less than a square mile.

That of the University of Iowa is on Okoboji Lake near Milford, Iowa. That of the University of Minnesota is on Lake Itasca, the source of the Mississippi River, in Itasca Park. That of the University of Virginia is at Mountain Lake (altitude 4000 ft.). That of Brigham Young University is on Utah Lake near Provo, Utah. That of the Tennessee Academy of Science is on Reelfoot Lake (a large shallow lake formed by an earthquake in 1811) near Memphis.

Under the direction of the Biological Board of Canada, which has its headquarters at the University of Toronto, much survey work is being done on Canadian lakes throughout the interior provinces, in cooperation with that University, with the provincial universities of Manitoba and Saskatchewan, and with Queen's University at Kingston. This work, like the survey work of the U. S. Bureau of Fisheries, is mainly done from temporary field stations, without the establishment of permanent laboratories.

Depth and Breadth—The depth of lakes is of more biological significance than the form of their basins; for, as we have seen in the preceding chapter, with increase of depth goes increased pressure, diminished light, and thermal stratification of the water. Living conditions are therefore very different in shallow water from what they are in the bottom of a deep lake, where there is no light, and where the temperature remains constant throughout the year. Absence of light prevents the growth of chlorophyl-bearing organisms and renders such waters relatively barren. The lighted top layer of the water (zone of photosynthesis) is the productive area. The other is a reservoir, tending to stabilize conditions. Lakes may therefore be roughly

grouped in two classes: first, those that are shallow
enough for complete circulation of their water by wind
or otherwise at any time; and second those deep enough
to maintain through a part of the summer season a
bottom reservoir of still water, undisturbed by waves or
currents, and stratified according to temperature and
consequent density. In these deeper lakes a thermocline appears during midsummer. In the lakes of New
York its upper limit is usually reached at about thirty-
five feet and it has an average thickness of some fifteen
feet. Our lakes of the second class may therefore be
said to have a depth greater than fifty feet.

Lakes of this class may differ much among themselves according to the relative volume of this bottom
reservoir of quiet water, Lakes Otisco and Skaneateles
(see map on page 65) serve well for comparison in this
regard, since they are similar in form and situation and
occupy parallel basins but a few miles apart.

Lake	Area in sq. mi.	Max. depth in ft.	% of vol. below 50 ft.	Trans- parency*	Free CO_2† at surface	bottom	Oxygen† at surface	bottom
Otisco	2.64	66	7.0	9.2	—2.50	+3.80	6.72	0.00
Skaneateles	13.90	297	70.2	31.8	—1.25	+1.00	6.75	7.89

*In feet, measured by Secchi's disc.

†In cc. per liter of water. Alkalinity by phenolthalein test is indicated
by the minus sign.

The figures given are from midsummer measurements by Birge and Juday. At the time these observations were made both lakes were alkaline at the surface.
tho still charged with free carbon dioxide at the bottom.
Apparently, the greater the body of deep water the
greater the reserve of oxygen taken up at the time of
the spring circulation and held through the summer
season. Deep lakes are as a rule less productive of
plancton in summer, even in their surface waters,
because their supply of available carbon dioxide runs
low. It is consumed by algae and carried to the bottom

with them when they die, and thus removed from circulation.

Increasing breadth of surface means increasing exposure to winds with better aeration, especially where waves break in foam and spray, and with the development of superficial currents. Currents in lakes are not controlled by wind alone, but are influenced as well by contours of basins, by outflow, and by the centrifugal pull due to the rotation of the earth on its axis. In Lake Superior a current parallels the shore, moving in a direction opposite to that of the hands of a clock. Only in the largest lakes are tides perceptible, but there are other fluctuations of level that are due to inequalities of barometric pressure over the surface. These are called *seiches*.

Broad lakes are well defined, for they build their own barrier reefs across every low spot in the shores, and round out their outlines. It is only shores that are not swept by heavy waves that merge insensibly into marshes. In winter in our latitude the margins of the larger lakes become icebound, and the shoreline is temporarily shifted into deeper water (compare summer and winter conditions at the head of Cayuga lake as shown in our frontispiece).

Increasing breadth has little effect on the life of the open water, and none, directly, on the inhabitants of the depths; but it profoundly affects the life of the shoals and the margins, where the waves beat, and the loose sands scour and the ice floes grind. Such a beach as that shown on page 61 is bare of vegetation only because it is storm swept. The higher plants cannot withstand the pounding of the waves and the grinding of the ice on such a shore.

The shallower a lake is the better its waters are exposed to light and air, and, other things being equal, the richer its production of organic life.

High and low water—Since the source of this water is in the clouds, all lakes fluctuate more or less with variation in rainfall. The great lakes drain an empire of 287,688 square miles, about a third of which is covered by their waters. They constitute the greatest system of fresh water reservoirs in the world, with an unparalleled uniformity of level and regularity of outflow. Yet their depth varies from month to month

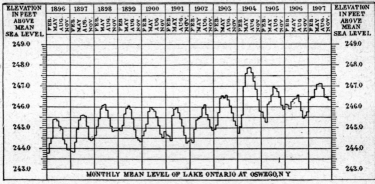

FIG. 19. Diagram of monthly water levels in Lake Ontario for twelve years, from the Report of the International Waterways Commission for 1910.

and from year to year, as shown on the accompanying diagram. From this condition of relative stability to that of regular disappearance, as of the strand lakes of the Southwest, there are all gradations. Topography determines where a lake may occur, but climate has much to do with its continuance. Lakes in arid regions often do not overflow their basins. Continuous evaporation under cloudless skies further aided by high winds, quickly removes the excess of the floods that run into them from surrounding mountains. The minerals dissolved in these waters are thus concentrated, and they become alkaline or salt. We shall have little to say in this book about such lakes, or about their population, but they constitute an interesting class. Life in their waters must meet conditions physiologically so different that few organisms can live in both fresh and salt water.

Large lakes in arid regions are continually salt; permanent lakes of smaller volume are made temporarily fresh or brackish by heavy inflowing floods; while

FIG. 20. Marl pond near Cortland, N. Y., at low water. The whiteness of the bed surrounding the residual pool is due to deposited marl, largely derived from decomposed snail shells. The marl is thinly overgrown with small freely-blooming plants of *Polygonum amphibium*. Tall aquatics mark the vernal shore line. (Photo by H. H. Knight).

strand lakes (called by the Spanish name *playa* lakes, in the Southwest) run the whole gamut of water content, and vanish utterly between seasons of rain.

Complete withdrawal of the waters is of course fatal to all aquatic organisms, save a few that have specialized means of resistance to the drought. Partial withdrawal

by evaporation means concentration of solids in solution, and crowding of organisms, with limitation of
their food and shelter. The shoreward population of
all lakes is subject to a succession of such vicissitudes.

The term *limnology* is often used in a restricted sense
as applying only to the study of freshwater lakes.
This is due to the profound influence of the Swiss
Master, F. A. Forel, who is often called the "Father
of Limnology." He was the first to study lakes
intensively after modern methods. He made the
Swiss lakes the best known of any in the world.
His greatest work "*Le Leman,*" a monograph on
Lake Geneva, is a masterpiece of limnological literature. It was he who first developed a comprehensive plan for the study of the life of lakes and all
its environing conditions.

STREAMS

OURNEYING seaward, the water that finds no basins to retain it, forms streams. According as these differ in size we call them rivers, creeks, brooks, and rills. These differ as do lakes in the dissolved contents of their waters, according to the nature of the soils they drain. Streams differ most from the lakes in that their waters are ever moving in one direction, and ever carrying more or less of a load of silt. From the geologist's point of view the work of rivers is the transportation of the substance of the uplands into the seas. It is an eternal levelling process. It is well advanced toward completion in the broad flood plains of the larger continental streams (see map on page 63); but only well begun where brooks and rills are invading the high hills, where the waters seek outlets in all directions, and where every slope is intersected with a maze of channels. The rapidity of the grading work depends chiefly upon climate and rainfall, on topography and altitude and on the character of the rocks and soil.

The rivers of America have been extensively studied
as to their hydrography, their navigability, their water-
power resources, and their liability to overflow with
consequent flood damage; but it is the conditions they

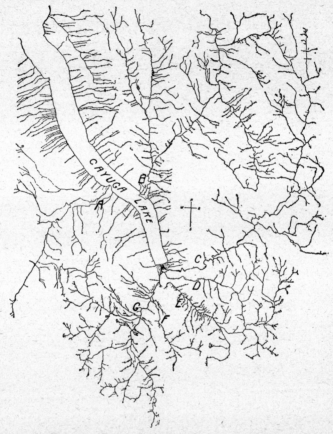

FIG. 21. Streams of the upper Cayuga basin.

A. Taughannock Creek, with a waterfall 211 feet high near its mouth;
B. Salmon Creek; *C*. Fall Creek with the Cornell University Biological Field
Station in the marsh at its mouth (views on this stream are shown in the initial
cuts on pages 24 and 82); *D*. Cascadilla Creek (view on page 55); *E*. Sixmile
Creek; *F*. Buttermilk Creek with Coys Glen opposite its mouth.   (View on
page 77; of the Glen on page 25); *G*. Neguena Creek or the Inlet.   The southernmost of these streams rise in cold swamps, which drain southward also into
tributaries of the Susquehanna River.

afford to their plant and animal inhabitants that
interest us here; and these have been little studied.
Most has been done on the Illinois River, at the floating
laboratory of the Illinois State Laboratory of Natural
History (see page 50). A more recently established
river laboratory, more limited in its scope (being
primarily concerned with the propagation of river
mussels) is that of the U. S. Fish Commission at Fairport, Iowa, on the Mis-
sissippi River.

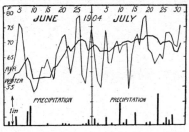

In large streams, espec-
ially in their deeper and
more quiet portions, the
conditions of life are most
like those in lakes. In lesser streams life is subject
to far greater vicissitudes.
The accompanying figure
shows comparative sum-
mer and winter tempera-
tures in air and in water of
Fall Creek at Ithaca. This
creek (see the figure on
page 24), being much
broken by waterfalls and
very shallow, shows hardly
any difference between surface and bottom tempera-

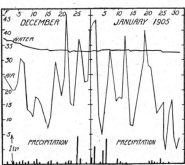

tures. The summer tem-
peratures of air and water
(fig. 22) are seen to maintain a sort of correspondence, in spite of the thermal

FIG. 22. Diagram showing summer
and winter conditions in Fall Creek
at Ithaca, N. Y. Data on air
temperatures furnished by Dr. W.M.
Wilson of the U. S. Weather Bureau.
Data on water temperatures by Professor E. M. Chamot.

conservatism of water, due to its greater specific heat.
This approximation is due to conditions in the creek
which make for rapid heating or cooling of the water.

It flows in thin sheets over broad ledges of dark colored rocks that are exposed to the sun, and it falls over projecting ledges in broad thin curtains, outspread in contact with the air.

The curves for the two winter months, show less concurrence, and it is strikingly apparent that during that period when the creek was ice-bound (Dec. 15–Jan. 31) the water temperature showed no relation to air temperature, but remained constantly at or very close to 0° C. (32° F.).

Forbes and Richardson (13) have shown how great may be the aerating effect of a single waterfall in such a sewage polluted stream as the upper Illinois River. "The fall over the Marseilles dam (710 feet long and 10 feet high) in the hot weather and low water period of July and August, 1911, has the effect to increase the dissolved oxygen more than four and a half times, raising it from an average of .64 parts per million to 2.94 parts. On the other hand, with the cold weather, high oxygen ratios, and higher water levels of February and March, 1912, and the consequent reduced fall of water at Marseilles, the oxygen increase was only 18 per cent.—from 7.35 parts per million above the dam to 8.65 parts below * * * The beneficial effect is greatest when it is most needed—when the pollution is most concentrated and when decomposition processes are most active."

Ice—The physical conditions that in temperate regions have most to do with the well- or ill-being of organisms living in running water are those resulting from the freezing. The hardships of winter may be very severe, especially in shallow streams. One may stand beside Fall Creek in early winter when the thin ice cakes heaped with snow are first cast forth on the stream, and see through the limpid water an abundant life gathered upon the stone ledges, above which these miniature floes are harmlessly drifting. There are great black patches of *Simulium* larvae, contrasting strongly with the whiteness of the snow. There are beautiful green drapings of *Cladophora* and rich red-purple fringes of *Chantransia*, and everywhere amber-brown carpetings of diatoms, overspreading all the bottom. But if one stand in the same spot in the spring, after the heavy ice of winter has gone out, he will see that the rocks have been swept clean and bare, every living thing that the ice could reach having gone.

The grinding power of heavy ice, and its pushing power when driven by waves or currents, are too well known to need any comment. The effects may be seen on any beach in spring, or by any large stream. But there is in brooks and turbulent streams a cutting with fine ice rubble that works through longer periods, and adds the finishing touches of destructiveness. It is driven by the water currents like sand in a blast, and it cleans out the little crevices that the heavy ice could not enter. This ice rubble is formed at the front of water falls under such conditions as are shown in the accompanying figure of Triphammer Falls at Ithaca. The pool below the fall froze first. The winter increasing cold, the spray began to freeze where it fell. It formed icicles, large and small, wherever it could find a support above. It built up grotesque columns on the edge of the ice of the pool beneath. It grew inward from the sides and began to overarch the stream face; and then, with favoring intense cold of some days duration, it extended these lines of frozen spray across the front of the fall in all directions, covering it as with a beautiful veil of ice.

The conditions shown in the picture are perfect for the rapid formation of ice rubble. From thousands of points on the underside of this tesselated structure minute icicles are forming and their tips are being broken off by the oscillations of the current. These broken tips constitute the rubble. They are sometimes remarkably uniform in size—those forming when this picture was taken were about the size of peas—and though small they are the tools with which the current does its winter cleaning. In the ponds formed by damming rapid streams this rubble accumulates under the solid ice.

"Anchor ice" forms in the beds of rapid streams, and adds another peril to their inhabitants. The water, cooled below the freezing point by contact with the air,

FIG. 23.   The ice veil on Triphammer Falls, Cornell University Campus. The fall is at the left, the Laboratory of Hydraulic Engineering at the right in the picture, the only open water seen is in the foaming pool at the foot of the fall.

does not freeze in the current because of its motion, but it does freeze on the bottom where the current is sufficiently retarded to allow it. It congeals in semi-solid or more or less flocculent masses which, when attached to the stones of the bed, often buoy them up

FIG. 24. A brook in winter. Country Club woods, Ithaca, N. Y.
Photo by John T. Needham.

and cause them to be carried away. Thus the organisms that dwell in the stream bed are deprived of their shelter and exposed to new perils.

Below the frost-line, however, in streams where dangers of mechanical injuries such as above mentioned are absent, milder moods prevail. In the bed of a gentle meandering streamlet like that shown in the accompanying figure, life doubtless runs on in winter with greater serenity than on land. Diatoms grow and caddis-worms forage and community life is actively maintained.

Silt—Part of the substance of the land is carried seaward in solution. It is ordinarily dissolved at or near the surface of the ground, but may be dissolved from underlying strata, as in the region of the Mammoth Cave in Kentucky, where great streams run far under ground. But the greater part is carried in suspension. Materials thus carried vary in size from the finest particles of clay to great trees dropped whole into the stream by an undercutting flood. The lighter solids float, and are apt to be heaped on shore by wave and wind. The heavier are carried and rolled along, more or less intermittently, hastened with floods and slackened with low water, but ever reaching lower levels. The rate of their settling in relation to size and to velocity of stream has been discussed in the preceding chapter.

Silt is most abundant at flood because of the greater velocity of the water at such times. Kofoid ('03) has studied the amount of silt carried by the Illinois River at Havana. Observations at one of his stations extending over an entire year show a minimum amount of 140 cc. per cubic meter of river water; a maximum of 4,284 cc., and an average for the year (28 samples) of 1,572 cc. Silt in a stream affects its population in a number of ways. It excludes light and so interferes with the growth of green plants, and thus indirectly with the food supply of animals. It interferes with the free locomotion of the microscopic animals by becoming entangled in their swimming appendages. It clogs the respiratory apparatus of other animals. It falls in deposits that smother and bury both plants and animals living on the bottom. Thus the best foraging grounds of some of our valuable fishes are ruined.

Professor Forbes ('oo) has shown that the fine silt from the earlier-glaciated and better weathered soils of southern Illinois, has been a probable cause of exclusion of a number of regional fishes from the streams of that portion of the state.

It is heavier silt that takes the larger share in the building of bars and embankments along the lower reaches of a great stream, in raising natural levees to hold impounded backwaters, and in blocking cut-off channels to make lakes of them.

Current—Rate of streamflow being determined largely by the gradient of the channel, is one of the more constant features of rivers, but even this is subject to considerable fluctuation according to volume. Kofoid states that water in the Illinois River travels from Utica to the mouth (227 miles) in five days at flood, but requires twenty-three days for the journey at lowest water. The increase in speed and in turbulence in flood time appears to have a deleterious effect upon some of the population, many dead or moribund individuals of free swimming entomostraca being present in the waters at such times.

With the runoff after abundant rainfall a rapid rise and acceleration occurs, to be followed by a much slower decline. The stuffs in the water are diluted; the plancton is scattered. A new load of silt is received from the land; plant growths are destroyed and even contours in the channel are shifted.

Current is promoted by increasing gradient of stream· bed. It is diminished by obstructions, such as rocks or plant growths, by sharp bends, etc. It is slightly accelerated or retarded by wind according as the direction is up or down stream. Even where a stream appears to be flowing steadily over an even bed between smooth shores, careful measurements reveal slight and inconstant fluctuations. The current is nowhere uniform from top to bottom or from bank to bank. In the horizontal plane it is swiftest in midstream and is retarded by the banks. In a vertical plane, it is swiftest just beneath the surface and is retarded more and more toward the bottom. The pull of the surface film retards it a little and when ice forms on the surface, friction against the ice retards it far more and throws the point of maximum velocity down near middepth of the stream. A sample measurement made by Mr. Wilbert A. Clemens in Cascadilla Creek at Ithaca in open water seventeen inches deep gave rate of flow varying from a maximum of 3.91 feet per second two inches below the surface down to 1.73 feet per second one inch above the bottom, as shown in the columns above. Below this, in the last inch of depth the retardation was more rapid, but irregular. The current slackens more slowly toward the surface and toward the side margins of the stream.

Depth in inches	Feet per sec.
2	3.91
3	3.73
4	3.60
5	3.32
6	3.04
7	2.89
8	2.81
10	2.73
12	2.64
14	2.46
15	2.17
16	1.73

Current and Depth in Cascadilla Creek. Measured by W. A. Clemens.

Mr. Clemens, using a small Pitot-tube current meter, made other measurements showing that in the places where dwell the majority of the inhabitants of swift streams there is much less current than one might expect. In the shelter of stones and other obstructions there is slack water. On sloping bare rock bottoms under a swiftly gliding stream the current is often but half that at the surface. On stones exposed to the current a coating of slime and diatomaceous ooze reduces the current 16 to 32 per cent.

This accounts for the continual restocking of a stream whose waters are swifter than the swimming of the animals found in the open channels. In these more or less shoreward places they breed and renew the supply. Except in a stream whose waters run a long course seaward, allowing an ample time for breeding, there is little indigenous free-swimming population.

FIG. 25. Annually inundated bulrush-covered flood-plain at the mouth of Fall Creek, Ithaca, N. Y., in 1908. Clear growth of *Scirpus fluviatilis* and a drowned elm tree. The Cornell University Biological Field Station at extreme right. West Hill in the distance.

High and Low Water—Inconstancy is a leading characteristic of river environment, and this has its chief cause in the bestowal of the rain. Streams fed mainly by springs, lakes, and reservoirs are relatively constant; but nearly all water courses are subject to overflow; their channels are not large enough to carry flood waters, so these overspread the adjacent bottomlands. Every change of level modifies the environment by connecting or cutting off back waters, by shifting currents, by disturbing the adjustment of the vegetation, and by causing the migration of the larger animals. At low water the Illinois River above Havana has a width of some 500 feet; in flood times it spreads across the valley floor in a sheet of water four miles wide.

The rise of a river flood is often sudden; the decline is always much more gradual, for impounding barriers and impeding vegetation tend to hold the water upon the lowlands. The period of inundation markedly affects the life of the land overflowed. Cycles of seasons with short periods of annual submergence favor the establishment of upland plants and trees. Cycles of years of more abundant rainfall favor the growth of swamp vegetation. Certain plants like the flood-plain bulrush shown in the preceding figure seem to thrive best under inconstancy of flood conditions.

A GREAT aquatic environment may be maintained with
much less water than there is in a lake or a river if only
an area of low gradient, lacking proper basin or channel,
be furnished with a ground cover of plants suitable for
retaining the water on the soil.   Enough water must be
retained to prevent the complete decay of the accumu-
lating plant remains.   Then we will have, according to
circumstances, a marsh, a swamp or a bog.

There are no hard and fast distinctions between these
three; but in general we may speak of a marsh as
a meadow-like area overgrown with herbaceous aquatic
plants, such as cat-tail, rushes and sedges; of a swamp
as a wet area overgrown with trees; and of a bog as such
an area overgrown with sphagnum or bog-moss, and
yielding under foot.   The great Montezuma Marsh of
Central New York (shown in the initial above) is

typical of the first class; the Dismal Swamp of eastern
Virginia, of the second; and over the northern lake
region of the continent there are innumerable examples
of the third. These types are rarely entirely isolated,
however, since both marsh and bog tend to be invaded
by tree growth at their margins. Such wet lands occupy
a superficial area larger by far than that covered by
lakes and rivers of every sort. They cover in all
probably more than a hundred million acres in the
United States; great swamp areas border the Gulf
of Mexico, the South Atlantic seaboard, and the lower
reaches of the Mississippi, and of its larger tributaries,
and partially overspread the lake regions of upper
Minnesota, Wisconsin, Michigan and Maine. In the
order of the areas of "swamp land" (officially so designated) within their borders the leading states are
Florida, Louisiana, Arkansas, Mississippi, Michigan,
Minnesota, Wisconsin and Maine.

Swamps naturally occupy the shoal areas along the
shores of lakes and seas. Marine swamps below mean
tide occur as shoals covered with pliant eel-grass.
Above mean tide they are meadow-like areas located
behind protecting barrier reefs, or they are mangrove
thickets that fringe the shore line, boldly confronting
the waves. With these we are not here concerned.
Fresh-water marshes likewise occupy the shoals bordering the larger lakes, where protected from the waves by
the bars that mark the shore line. In smaller lakes,
where not stopped by wave action, they slowly invade
the shoaler waters, advancing with the filling of the
basin, and themselves aiding in the filling process.

That erosion sometimes gives rise to lakes has
already been pointed out; much oftener it produces
marshes; for depositions of silt in the low reaches of
streams are much more likely to produce shoals than
deep water.

Cat-tail Marshes—In the region of great lakes every open area of water up to ten feet in depth is likely to be invaded by the cat-tail flag (*Typha*). The ready dispersal of the seeds by winds scatters the species everywhere, and no permanent wet spot on the remotest hill-top is too small to have at least a few plants. Along

FIG. 26. An open-water area (Parker's Pond) in the Montezuma Marsh in Central New York. Formerly it teemed with wild water fowl. It is surrounded by miles of cat-tail flags (Typha) of the densest sort of growth.

the shores of the Great Lakes and in the broad shoals bordering on the Seneca River there are meadow-like expanses of Typha stretching away as far as the eye can see. Many other plants are there also, as will be noted in a subsequent chapter, but Typha is the dominant plant, and the one that occupies the fore-front of the advancing shore vegetation. It masses its crowns and numberless interlaced roots at the surface of the water in floating rafts, which steadily extend into deeper water. The pond in the center of Montezuma Marsh shown in the preceding figure is completely surrounded by a rapidly advancing, half-floating even-fronted phalanx of cat-tail.

FIG. 27. "The Cove" at the Cornell University Biological Field Station, in time of high water. Early summer. Two of the University buildings appear on the hill in the distance.

Later conditions in such a marsh are those illustrated by our frontispiece: regularly alternating spring floods, summer luxuriance, autumn burning and winter freezing. This goes on long after the work of the cat-tail, the pioneer landbuilding, has been accomplished. The excellent aquatic collecting ground shown in the accompanying figure is kept open only by the annual removal of the encroaching flag.

The Okefenokee Swamp.　In southern Georgia lies this most interesting of American swamps. It is formed behind a low barrier that lies in a N., N. E.— S., S. W. direction across the broad sandy coastal plain, intersecting the course of the southernmost rivers of

Fig. 28. A view of "Chase's Prairie" in the more open eastern portion of the Okefenokee Swamp, taken from an elevation of fifty feet up a pine tree on one of the incipient islets. The water is of uniform depth (about four or five feet). This is one of the most remarkable landscapes in the world.
Photo by Mr. Francis Harper.

Georgia that drain into the Atlantic. Behind the barrier the waters coming from the northward are retained upon a low, nearly level plain, that is thinly overspread with sand and underlaid with clay. They cover an area some forty miles in diameter, hardly anywhere too deep for growth of vascular plants. There is little discoverable current except in the nascent channels of the two outflowing streams, St. Mary's and Suwannee Rivers. The waters are deeper over the eastern part of the swamp, the side next the barrier; and here the vegetation is mainly herbaceous plants, principally submerged aquatics, with occasional broad meadow-like areas overgrown with sedges. These are the so-called "prairies." The western part of the swamp (omitting from consideration the islands) is a true swamp in appearance being covered with trees, principally cypress. A few small strips of more open and deeper water (attaining 25 feet) of unique beauty, owing to their limpid brown waters and their setting of Tillandsia-covered forest, are called lakes.

The whole swamp is in reality one vast bog. Its waters are nearly everywhere filled with sphagnum. Whatever appears above water to catch the eye of the traveler, whether cypress and tupelo in the western part or sedges and water lilies on the "prairie," everywhere beneath and at the surface of the water there is sphagnum; and it is doubtless to the waterholding capacity of this moss that the relative constancy of this great swamp on a gently inclined plain near the edge of the tropics, is due.

Climbing bogs—In-so-far as swamps possess any basin at all they approximate in character to shallow lakes; but there are extensive bogs in northern latitudes that are built entirely on sloping ground; often even on convex slopes. These are the so-called "climbing bogs." They belong to cool-temperate and humid regions. They exist by the power of certain plants, notably sphagnum, to hold water in masses, while giving off very little by evaporation from the surface. A climbing bog proceeds slowly to cover a slope by the growth of the mass of living moss upward against gravity, and in time what was a barren incline becomes a deep spongy mass of water soaked vegetation.

Conditions of life—In the shoal vegetation-choked waters of marshes there is little chance for the formation of currents and little possibility of disturbance by wind. Temperature conditions change rapidly, however, owing to the heat absorbing and heat radiating power of the black plant-residue. The diurnal range is very great, water that is cool of a morning becomes repellantly hot of a summer afternoon. Temperatures above 90° F. are not then uncommon. Unpublished observations made by Dr. A. A. Allen in shoal marsh ponds at the Cornell Biological Field Station throughout the year 1909, show a lower temperature at the surface of the water than in the bottom mud from December to April, with reverse conditions for the remainder of the year. The black mud absorbs and radiates heat rapidly.

Conditions peculiar to marshes, swamps and bogs are those due to massed plant remains more or less permanently saturated with water. Water excludes the air and hinders decay. Half disintegrated plant fragments accumulate where they fall, and continue for a longer or shorter time unchanged. According to their state of decomposition they form peat or muck.

In peat the hard parts and cellular structure of the plant are so well preserved that the component species may be recognized on microscopic examination. To the naked eye broken stems and leaves appear among the finer fragments, the whole forming a springy or spongy mass of a loose texture and brownish color. The color deepens with age, being lightest immediately under the green and living vegetation, and darkest in the lower strata, where always less well preserved.

The water that covers beds of peat acquires a brownish color and more or less astringent taste due to infusions of plant-stuffs. Humous acids are present in abundance and often solutions of iron sulphate and other minerals.

Muck is formed by the more complete decay of such water plants as compose peat. The process of decay is furthered either by occasional exposure of the beds to the air in spells of drought, or by the presence of lime in the surrounding soil, correcting the acidity of the water and lessening its efficiency as a preservative. Muck is soft and oozy, paste-like in texture and black in color. In the openings of marshes, like that shown on page 89 are beds of muck so soft that he who ventures to step on it may sink in it up to his neck. In such a bed the slow decomposition that goes on in hot weather in absence of oxygen produces gases that gather in bubbles increasing in size until they are able to rise and disrupt the surface.* So are formed marsh gas (methane) which occasionally ignites spontaneously, in mysterious flashes over the water—the well known "Jack-o-lantern" or "Will-o-the-wisp" or *"Ignis fatuus"*—and hydrogen sulphide which befouls the surrounding atmosphere.

The presence in marsh pools of these noxious gases, of humous acids, and of bitter salts, and of the absence of oxygen—except at the surface, limits their animal population in the main to such creatures as breathe air at the surface or have specialized means of meeting these untoward conditions.

High and Low Water—Swamps being the shoalest of waters are subject to the most extreme fluctuations. That they retain through most dry seasons enough water for a permanent aquatic environment is largely due to the water-retaining power of aquatic plants. Notable among these is sphagnum, which holds enmeshed in its leaves considerable quantities of water, lifted above the surrounding water level. Aquatic seed

---

*See Penhallow, "A blazing beach" in *Science*, 22:794–6, 1905.

plants, also, whose stems in life are occupied with capacious air spaces, fill with water when dead and fallen, and hold it by capillarity. So, they too, form in partial decay a soft spongy water-soaked ground cover.

Marshes develop often a wonderful density of population, for they have at times every advantage of water, warmth and light. The species are fewer, however, than in the more varied environment of land. Comparatively few species are able to maintain themselves permanently where the pressure for room is so great when conditions for growth are favorable, and where these conditions fail more or less completely every dry season. Aquatic creatures that can endure the conditions shown in the accompanying figure must have specialized means of tiding over the period of drouth.

FIG. 29.   The bed of a marsh pool in a dry season, showing deep mud cracks, and a thin growth of bur-marigold and spike-rush.

A DIAGRAM OF LIGHT DISAPPEARANCE IN LAKE GEORGE, N. Y.

From a Report by the senior author. By
Courtesy of the New York Conservation Commission

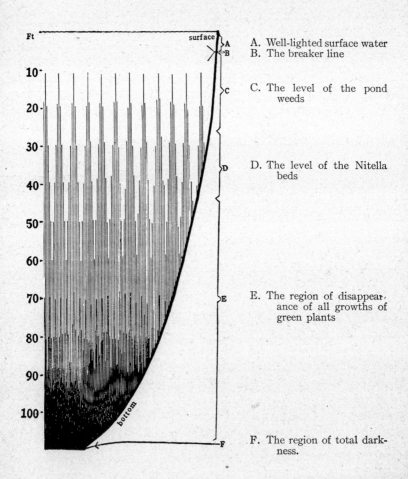

A. Well-lighted surface water
B. The breaker line

C. The level of the pond weeds

D. The level of the Nitella beds

E. The region of disappearance of all growths of green plants

F. The region of total darkness.

CHAPTER IV

AQUATIC ORGANISMS

IS the testimony of all biology that the water was the original home of life upon the earth. Conditions of living are simpler there than on the land. Food tends to be more uniformly distributed. The perils of evaporation are absent. Water is a denser medium than air, and supports the body better, and there is, in the beginning, less need of wood or bone or shell or other supporting structures. Life began in the water, and the simpler forms of both plants and animals are found there still.

But not all aquatic forms have remained simple. For when they multiplied and spread and filled all the waters of the earth the struggle for existence wrought diversification and specialization among them, in water as on land. The aquatic population is, therefore, a mixture of forms structurally of high and low degree. All the types of plant and animal organization are represented in it. But they are fitted to conditions so different from those under which terrestrial beings live as to seem like another world of life.

The population of the water includes besides the original inhabitants—those tribes that have always lived in the water—a mixture of forms descended from

ancestors that once lived on land. The more primitive groups are most persistently aquatic. Comparatively few members of those groups that have become thoroughly fitted for life on land have returned to the water to live.

WATER PLANTS

VERY large group of plants has its aquatic members. The algae alone are predominantly aquatic. Most of them live wholly immersed; some live in moist places, and a few in dry places, having special fitnesses for avoiding evaporation. In striking contrast with this, all the higher plants, the seed plants, ferns, and mosses, center upon the land, having few species in wet places and still fewer wholly immersed. Their heritage of parts specially adapted to life on land is of little value in the water. Rhizoids as foraging organs, a thick epidermis with automatic air pores, and strong supporting tissues are little needed under water. These plants have all a shoreward distribution, and do not belong to the open water. Only algae, molds and bacteria are found in all waters.

THE ALGAE

It is a vast assemblage of plants that makes up this group; and they are wonderfully diverse. Most of them are of microscopic size, and few of even the larger ones intrude upon our notice. Notwithstanding their elegance of form, their beauty of coloration and their great importance in the economy of water life, few of them are well known. However, certain mass effects produced by algae are more or less familiar. Massed together in inconceivably vast numbers upon the surface of still water, their microscopic hosts compose the "water bloom." Floating free beneath the surface they give to the water tints of emerald* of amber† or of blood‡. Matted masses of slender green filaments compose the growths that float on oxygen bubbles to the surface in the spring as "pond scums." Lesser masses of delicately branched filaments fringe the rocky ledges in the path of the cataract, or encircle submerged sticks and piling in still waters. Mixtures of various gelatinous algae coat the flat rocks in clear streams, making them green and slippery; and a rich amber-tinted layer of diatom ooze often overspreads the stream bed in clear waters.

These are all mass effects. To know the plants composing the masses one must seek them out and study them with the microscope. Among all the hosts of fresh water algae, only a few of the stoneworts (Characeae) are in form and size comparable with the higher plants.

Many algae are unicellular; more are loose aggregates of cells functioning independently; a few are well integrated bodies of mutually dependent cells.

---

*Volvox in autumn in waters over submerged meadows of water weed.
†Dinobryon in spring in shallow ponds.
‡*Trichodesmium erythræum* gives to the Red Sea the tint to which its name is due. The little crustacean, Diaptomus, often gives a reddish tint to woodland pools.

The cells sometimes form irregular masses, with more or less gelatinous investiture. Often they form simple threads or filaments, or flat rafts, or hollow spheres. Algal filaments are sometimes simple, sometimes branched; sometimes they are cylindric, sometimes tapering; sometimes they are attached and grow at the free end only; sometimes they grow throughout; sometimes they are free, sometimes wholly enveloped in transparent gelatinous envelopes. And the form of the ends, the sculpturing and ornamentation of the walls and the distribution of chlorophyll and other pigments are various beyond all enumerating, and often beautiful beyond description. We shall attempt no more, therefore, in these pages than a very brief account of a few of the commoner forms, such as the general student of fresh water life is sure to encounter; these we will call by their common names, in so far as such names are available.

The flagellates—We will begin with this group of synthetic forms, most of which are of microscopic size and many of which are exceedingly minute. That some of them are considered to be animals (Mastigophora) need not deter us from considering them all together, suiting our method to our convenience. The group overspreads the undetermined borderland between plant and animal kingdoms. Certain of its members (*Euglena*) appear at times to live the life of a green plant, feeding on mineral solutions and getting energy from the sunlight; at other times, to feed on organic substances and solids like animals. The more common forms live as do the algae. All the members of the group are characterized by the possession of one or more living protoplasmic swimming appendages, called flagella, whence the group name. Each flagellum is long, slender and transparent, and often difficult of observation, even when the jerky movements of the attached cell give evidence of its presence and its activity. It swings in front of the cell in long serpentine curves, and draws the cell after it as a boy's arms draw his body along in swimming.

Many flagellates are permanently unicellular; others remain associated after repeated divisions, forming colonies of various forms, some of which will be shown in accompanying figures.

Carteria—This is a very minute flagellate of spherical form and bright green in color (fig. 30*a*). It differs from other green flagellates in having four flagella: the others have not more than two. It is widely distributed in inland waters, where it usually becomes more abundant in autumn, and it appears to prefer slow streams. Kofoid's notes concerning a maximum occurrence in the Illinois River are well worth quoting:

"The remarkable outbreak of Carteria in the autumn of 1907 was associated with unusually low water, and

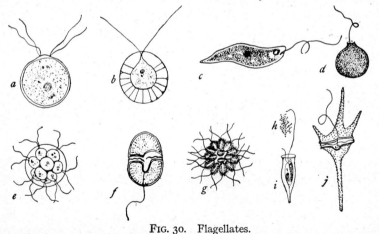

FIG. 30. Flagellates.

a, Carteria; b, Sphærella; c, Euglena; d, Trachelomonas; e, Pandorina; f, Glenodinium; g, Synura; h, and *i, Dinobryon;* a colony as it appears under low power of the microscope and a single individual highly magnified; *j, Ceratium,* (Reversed left for right in copying.)

concentration of sewage, and decrease of current. The water of the stream was of a livid greenish yellow tinge. * * * The distribution of Carteria in the river was remarkable. It formed great bands or streaks visible near the surface, or masses which in form simulated cloud effects. The distribution was plainly uneven, giving a banded or mottled appearance to the stream. The bands, 10 to 15 meters in width, ran with the channel or current, and their position and form were plainly influenced by these factors. No cause was apparent for the mottled regions. This phenomenon stands in somewhat sharp contrast with the usual distribution of waterbloom upon the river, which is generally composed largely of Euglena. This presents a much more uniform distribution, and unlike Carteria, is plainly visible only when it is accumulated as a superficial scum or film. Carteria was present in such quantity that its distribution was evident at lower levels so far as the turbidity would permit it to be seen. It afforded a striking instance of marked inequalities in distribution.''

Similar green flagellates of wide distribution are Chlamydomonas and Sphaerella (fig. 30*b*) commonly found in rainwater pools.

Certain aggregates of such cells into colonies are very beautiful and interesting. Small groups of such green cells are held together in flat clusters in Gonium and Platydorina, or in a hollow sphere, with radiating flagella that beat harmoniously to produce a regular rolling locomotion in Pandorina (fig. 30 *e*), Eudorina and Volvox.

Volvox—The largest and best integrated of these spherical colonies is Volvox (fig. 31). Each colony may consist of many thousands of cells, forming a sphere that is readily visible to the unaided eye. It rotates constantly about one axis, and moves forward therefore through the water in a perfectly definite manner. Moreover, the "eye spots" or pigment flecks of the individual cells are larger on the surface that goes fore-

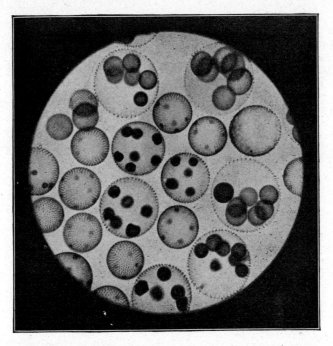

FIG. 31. Volvox, showing young colonies in all stages of development.

most. Sex cells are fully differentiated from the ordinary body cells. Nevertheless, new colonies are ordinarily reproduced asexually. They develop from single cells of the old colony which slip inward somewhat below the general level of the body cells, repeatedly divide, (the mass assuming spherical form), differentiate a full complement of flagella, a pair to each cell, and then escape to the outer world by rupturing the gelatinous walls of the old colony. Many developing colonies are shown within the walls of the old ones in the figure.

Often, when a weed-carpeted pond shows a tint of bright transparent green in autumn, a glass of the water, dipped and held to the light, will be seen to be filled with these rolling emerald spheres.

Euglena—Several species of this genus (fig. 30*c*) are common inhabitants of slow streams and pools. They are all most abundant in mid-summer, being apparently attuned to high temperatures. They are common constituents of the water-bloom that forms on the surface of slow streams. Figure 1 (p. 15) shows such a situation, where they recur every year in June. Certain of them are common in pools at sewer outlets, where bloodworms dwell in the bottom mud. When abundant in such places they give to the water a livid green color. Their great abundance makes them important agents in converting the soluble stuffs of the water into food for rotifers and other microscopic animals.

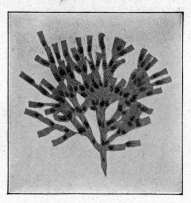

FIG. 32. A Dinobryon colony.

Dinobryon—This minute, amber-tinted flagellate forms colonies on so unique a plan (fig. 30*h*) they are not readily mistaken for anything else under the sun. Each individual is enclosed in an ovoid conic case or lorica, open at the front where two flagella protrude (fig. 30*i*) and many of them are united together in branching, a more or less tree-like colony. Since flagella

always draw the body after them, these colonies swim
along with open ends forward, apparently in defiance
of all the laws of hydromechanics, rotating slowly on
the longitudinal axis of the colony as they go. Dinobryon is of an amber yellow tint, and often occurs in
such numbers as to lend the same tint to the water it
inhabits. It attains its maximum development at
low temperatures. In the cooler waters of our larger
lakes it is present in some numbers throughout the year,
though more abundant in winter. Kofoid reports it as
being "sharply limited to the period from November to
June" in Illinois River waters. Its sudden increase
there at times in the winter is well illustrated by the
pulse of 1899, when the numbers of individuals per
cubic meter of water in the Illinois River were on successive dates as follows:

Jan. 10th, 1,500
Feb. 7th,   6,458,000
"   14th, 22,641,440

followed by a decline, with rising of the river.

Dinobryon often develops abundantly under the ice.
Its optimum temperature appears to be near 0° C. It
thus takes the place in the economy of the waters that
is filled during the summer by the smaller green
flagellates.

Synura (fig. 30g) is another winter flagellate, similar
in color and associated with Dinobryon, much larger
in size. Its cells are grouped in spherical colonies
united at the center of the sphere, and equipped on the
outer ends of each with a pair of flagella, which keep the
sphere in rolling locomotion. The colonies appear at
times of maximum development to be easily disrupted,
and single cells and small clusters of cells are often found
along with well formed colonies. Synura when abundant often gives to reservoir waters an odor of cucumbers,

and a singularly persistent bad flavor, and under such circumstances it becomes a pest in water supplies.

Glenodinium (fig. 30*f*), Peridinium, and Ceratium (fig. 30*j*) are three brownish shell-bearing flagellates of wide distribution often locally abundant, especially in spring and summer.   These all have one of the two long flagella laid in a transverse groove encircling the body, the other flagellum free (fig. 33).

Glenodinium is the smallest, Ceratium, much the largest. Glenodinium has a smooth shell, save for the grooves where the flagella arise.   Peridinium has a brownish chitinous shell, divided into finely reticulate plates.   Ceratium has a heavy grayish shell prolonged into several horns.

FIG. 33.   Ceratium (The transverse groove shows plainly, but neither flagellum shows in the photograph.)

On several occasions in spring we have seen the waters of the Gym Pond on the Campus at Lake Forest College as brown as strong tea with a nearly pure culture of Peridinium and concurrently therewith we have seen the transparent phantom-larvæ of the midge *Corethra* in the same pond all showing a conspicuous brown line where the alimentary canal runs through the body, this being packed full of Peridinia.

Trachelomonas (fig. 30 *d*) is a spherical flagellate having a brownish shell with a short flask-like neck at one side whence issues a single flagellum.   This we have found abundantly in pools that were rich in oak leaf infusions.

Diatoms—Diatoms are among the most abundant of living things in all the waters of the earth. They occur singly and free, or attached by gelatinous stalks, or

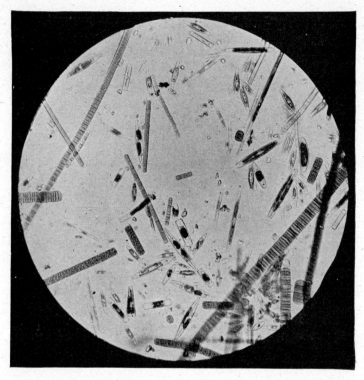

FIG. 34. Miscellaneous diatoms, mostly species of *Navicula;* the filaments are blue-green algae, mostly *Oscillatoria.*

aggregated together in gelatinous tubes, or compactly grouped in more or less coherent filaments. All are of microscopic size. They are most easily recognized by their possession of a box-like shell, composed of two *valves*, with overlapping edges. These valves are stiffened by silica which is deposited in their outer walls, often in beautiful patterns. The opposed edges of the

valves are connected by a membraneous portion of the
cell wall known as the *girdle*. A diatom may appear
very different viewed from the face of the valve, or from
the girdle (see fig. 35*a* and *b*, or *j* and *k*). They are
circular-like pill-boxes in one great group, and more or
less elongate and bilateral in the others.

Diatoms are rarely green in color. The chlorophyll
in them is suffused by a peculiar yellowish pigment
known as *diatomin*, and their masses present tints of
amber, of ochre, or of brown; sometimes in masses they
appear almost black. The shells are colorless; and,
being composed of nearly pure silica, they are well
nigh indestructible. They are found abundantly in
guano, having passed successively through the stomachs
of marine invertebrates that have been eaten by fishes,
that have been eaten by the birds responsible for the
guano deposits, and having repeatedly resisted diges-
tion and all the weathering and other corroding effects
of time. They abound as fossils. Vast deposits of
them compose the diatomaceous earths. A well-known
bed at Richmond, Va., is thirty feet in thickness and of
vast extent. Certain more recently discovered beds
in the Rocky Mountains attain a depth of 300 feet.
Ehrenberg estimated that such a deposit at Biln in
Bohemia contained 40,000,000 diatom shells per cubic
inch.

Singly they are insignificant, but collectively they are
very important, by reason of their rapid rate of increase,
and their ability to grow in all waters and at all ordinary
temperatures. Among the primary food gatherers
of the water world there is no group of greater importance.

In figure 35 we present more or less diagrammatically
a few of the commoner forms. The boat-shaped, freely
moving cells of *Navicula* (*a*, *b*, *c*) are found in every
pool. One can scarcely mount a tuft of algae, a leaf

of water moss or a drop of sediment from the bottom without finding Naviculas in the mount. They are more abundant shoreward than in the open waters of the lake. The "white-cross diatom" *Stauroneis* (*d*), is a kindred form, easily recognizable by the smooth cross-band which replaces the middle nodule of Navicula.

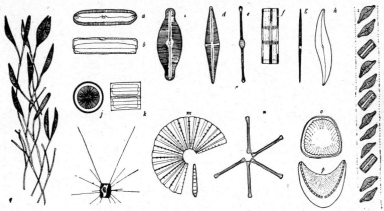

FIG. 35. Diatoms.

a, valve view showing middle and end nodules, and *b*, girdle view of *Navicula*. *c*, another species of *Navicula*; *d*, *Stauroneis*; *e*, valve view and *f*, girdle view of *Tabellaria*; *g*, *Synedra*; *h*, *Gyrosigma*; *i*, a gelatinous cord-like cluster of *Encyonema* showing girdle view of nine individuals and valve view of three. *j*, valve view and *k*, girdle view of *Melosira*; *l*, *Stephanodiscus*; *m*, *Meridion* colony, with a single detached individual shown in valve view below; *n*, a small colony of *Asterionella*; *o*, valve view, and *p*, girdle view of *Camplylodiscus*; *q*, cluster of *Cocconema*. (Figures mostly after Wolle).

Tabellaria (*e* and *f*) is a thin flat-celled diatom that forms ribbon-like bands, the cells being apposed, valve to valve. Often the ribbons are broken into rectangular blocks of cells which hang together in zig-zag lines by the corners of the rectangles. The single cell is long-rectangular in girdle view (slightly swollen in the middle and at each end, as shown at *e*, in valve view), and is traversed by two or more intermediate septa. Tabellaria abounds in the cool waters of our deeper northern lakes, at all seasons of the year. It is much less common in streams.

The slender cells of the "needle diatoms," *Synedra* (*g*), are common in nearly all waters and at all seasons. They are perhaps most conspicuously abundant when found, as often happens, covering the branches of some tufted algae, such as Cladophora, in loose tufts and fascicles, all attached by one end.

Gyrosigma (*h*) is nearly allied to Navicula but is easily recognized by the gracefully curved outlines of its more or less S-shaped shell. The sculpturing of this shell (not shown in the figure) is so fine it has long been a classic test-object for the resolving power of microscopic lenses. Gyrosigma is a littoral associate of Navicula, but of much less frequent occurrence.

Encyonema (*i*) is noteworthy for its habit of developing in long unbranched gelatinous tubes. Sometimes these tubes trail from stones on the bottom in swift streams. Sometimes they radiate like delicate filmy hairs from the surfaces of submerged twigs in still water. The tubes of midge larvae shown in figure 134 were encircled by long hyaline fringes of Encyonema filaments, which constituted the chief forage of the larvae in the tubes and which were regrown rapidly after successive grazings. When old, the cells escape from the gelatine and are found singly.

The group of diatoms having circular shells with radially arranged sculpturing upon the valves is represented by *Melosira* (*j* and *k*) and *Stephanodiscus* (*l*) of our figure. Melosira forms cylindric filaments, whose constituent cells are more solidly coherent than in other diatoms. Transverse division of the cells increases the length of the filaments, but they break with the movement of the water into short lengths of usually about half a dozen cells. They are common in the open water of lakes and streams, and are most abundant at the higher temperatures of midsummer. *Cyclotella* is a similar form that does not, as a rule, form filaments.

Its cells are very small, and easily overlooked, since they largely escape the finest nets and are only to be

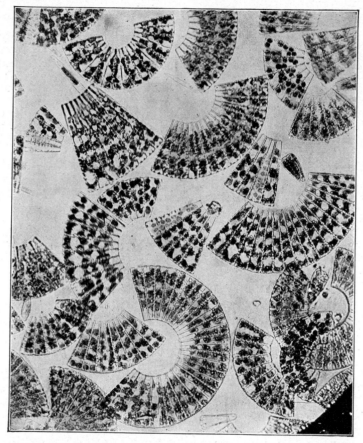

FIG. 36. A nearly pure culture of Meridion, showing colonies of various sizes.

gathered from the water by filtering. Often, however, their abundance compensates for their size. Kofoid found their average number in the waters of the Illinois

River to be 36,558,462 per cubic meter of water, and he considered them as one of the principal sources of food supply of Entomostraca and other microscopic aquatic animals. *Stephanodiscus* (*l*) is distinguished by the long, hyaline filaments that radiate from the ends of the box, and that serve to keep it in the water. A species of Stephanodiscus having shorter and more numerous filaments is common in the open waters of Cayuga Lake in spring.

The cells of *Meridion* are wedge-shaped, and grouped together side by side, they form a flat spiral ribbon of very variable length, sometimes in one or more complete turns, but oftener broken into small segments. This form abounds in the brook beds about Ithaca, covering them every winter with an amber-tinted or brownish ooze, often of considerable thickness. It appears to thrive best when the temperature of the water is near 0° C. Its richest growth is apparent after the ice leaves the brooks in the spring. As a source of winter food for the lesser brook-dwelling animals, it is doubtless of great importance. A view of a magnified bit of the ooze is shown in figure 36.

The colonies of *Asterionella* (*n*) whose cells, adhering at a single point, radiate like the spokes of a wheel, are common in the open waters of all our lakes and large streams. It is a common associate of Cyclotella, and of Tabellaria and other band-forming species, and is often more abundant than any of these. The open waters of Lake Michigan and of Cayuga Lake are often yellowish tinted because of its abundance in them. Late spring and fall (especially the former) after the thermal overturn and complete circulation of the water are the seasons of its maximum development. Asterionella abounds in water reservoirs, where, at its maxima, it sometimes causes trouble by imparting to the water an aromatic or even a decidedly "fishy" odor and an unpleasant taste.

Campylodiscus (*o* and *p*) is a saddle-shaped diatom of rather local distribution. It is found abundantly in the ooze overspreading the black muck bottom of shallow streams at the outlet of bogs. In such places in the upper reaches of the tributaries of Fall Creek near Ithaca it is so abundant as to constitute a large part of the food of a number of denizens of the bottom mud—notably of midge larvae, and of nymphs of the big Mayfly, Hexagenia.

These are a few—a very few—of the more important or more easily recognized diatoms. Many others will be encountered anywhere, the littoral forms especially being legion. Stalked forms like *Cocconema* (fig. 35*q* and fig. 37) will be found attached to every solid support. And minute close-clinging epiphytic diatoms, like *Cocconeis* and *Epithemia* will be found thickly besprinkling the green branches of many submerged aquatics. These adhere closely by the flat surface of one valve to the epidermis of aquatic mosses.

FIG. 37. A stalked colony *Cocconema.*

In open lakes, also, there are other forms of great importance, such as *Diatoma*, *Fragillaria*, etc., growing in flat ribbons, as does Tabellaria. It is much to be regretted that there are, as yet, no readily available popular guides to the study of a group, so important and so interesting. Equipped with a plancton net and a good microscope, the student would never lack for material or for problems of fascinating interest.

Desmids—This is a group of singularly beautiful unicellular fresh-water algae. Desmids are, as a rule, of a refreshing green color, and their symmetry of form and delicacy of sculpturing are so beautiful that they have always been in favor with microscopists. So

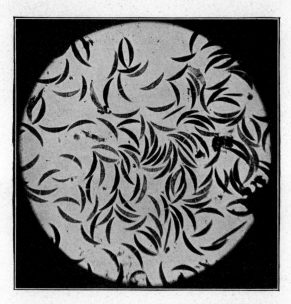

FIG. 38. A good slide-mount from a Closterium culture as it appears under a pocket lens. Two species.

numerous are they that their treatment has of late been relegated to special works. Here we can give only a few words concerning them, with illustrations of some of the commoner forms.

Desmids may be recognized by the presence of a clear band across the middle of each cell, (often emphasized by a corresponding median constriction) dividing it symmetrically into two semicells. Superficially they appear bicellular (especially in such forms as *Cylindro-*

cystis, fig. 40 *e*), but there is a single nucleus, and it lies in the midst of the transparent crossband. The larger ones, such as *Closterium* (fig. 38) may be recognized with the unaided eye, and may be seen clearly with a pocket lens. Because it will grow perennially in a culture jar in a half-lighted window, Closterium is a very well known laboratory type.

Division is transverse and separates between the semicells. Its progress in Closterium is shown in figure 39, in a series of successive stages that were photographed between 10 P. M. and 3 A. M. Division normally occurs only at night.

In a few genera (*Gonatozygon*, (fig. 40*a*) *Desmidium*, etc.) the cells after division remain attached, forming filaments.

Desmids are mainly free floating and grow best in still waters. They abound in northern lakes and peat bogs. They prefer the waters that run off archaean rocks and few of them flourish in waters rich in lime. A few occur on mosses in the edges of waterfalls, being attached to the

FIG. 39. Photomicrographs of a Closterium dividing. The lowermost figure is one of the newly formed daughter cells, not yet fully shaped.

mosses by a somewhat tenacious gelatinous investment. One can usually obtain a fine variety of desmids by squeezing wisps of such water plants as Utricularia and Sphagnum, over the edge of a dish, and examining the run-off. The largest genus of the group and also one of the most widespread is

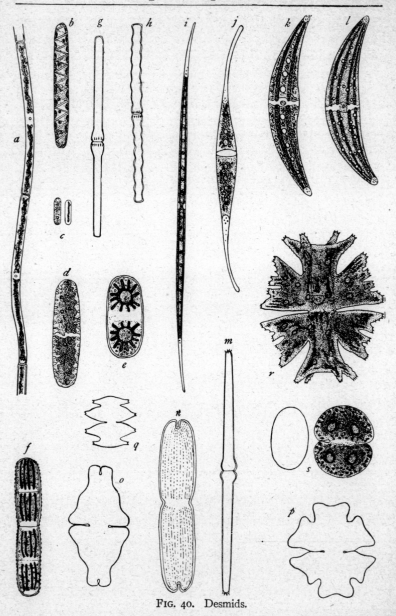

Fig. 40. Desmids.

Cosmarium (fig. 40 *s*). The most bizarre forms are found in the genera *Micrasterias* (figs. 40 *q* and *r*) and *Staurastrum*.

These connect in form through *Euastrum* (fig. 40 *o*) *Tetmemorus* (fig. 40 *n*) *Netrium* (fig. 40 *d*), etc., with the simpler forms which have little differentiation of the poles of the cell; and these, especially *Spirotaenia* (fig. 40 *b*) and *Gonatozygon* (fig. 40 *a*) connect with the filamentous forms next to be discussed.

The Filamentous Conjugates —This is the group of filamentous algae most closely allied with the desmids. It includes three common genera (fig. 41)—Spirogyra, Zygnema, and Mougeotia. The first of these being one of the most widely used of biological "types" is known to almost every laboratory student. Its long, green, unbranched, slippery filaments are easily recognized among all the other greenery of the water by their beautiful spirally-wound bands of chlorophyl. The other common genera have also distinctive chlorophyl arrangement. Zygnema has a pair of more or less star-shaped green masses in each cell, one on either side of the central nucleus. In Mougeotia the chlorophyl

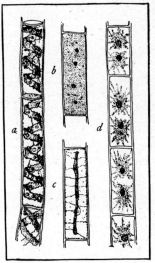

Fig. 41. Filamentous conjugates.

a. Spirogyra; b, flat view, and *c,* edgewise view of the chlorophyl plate in cells of *Mougeotia; d, Zygnema.*

a, a little more than two cells from a filament of *Gonatozygon*
b *Spirotænia*
c *Mesotænium*
d *Netrium*
e *Cylindrocystis*
f *Penium*
g *Docidium baculum*
h *Docidium undulatum*
i *Closterium pronum*
j *Closterium rostratum*
k *Closterium moniliferum*
l *Closterium ehrenbergi*
m *Pleurotænium*
n *Tetmemorus*
o *Euastrum didelta*
p *Euastrum verrucosum*
q *Micrasterias oscitans*
r *Micrasterias americana;* (for a third species see page 53).
s *Cosmarium,* face view, and outline as seen from the side

is in a median longitudinal plate, which can rotate in the cell: it turns its thin edge upward to the sun, but lies broadside exposed to weak light. Spirogyra is the most abundant, especially in early spring where it is found in the pools ere the ice has gone out. All, being unattached (save as they become entangled with rooted aquatics near shore), prefer quiet waters. Immense accumulations of their tangled filaments often occur on the shores of shallow lakes and ponds, and with the advance of spring and subsidence of the water level, these are left stranded upon the shores. They chiefly compose the "blanket-moss" of the fishermen. They settle upon and smother the shore vegetation, and in their decay they sometimes give off bad odors. Sometimes they are heaped in windrows on shelving beaches, and left to decay.

We most commonly see them floating at the surface in clear, quiet, spring-fed waters in broad filmy masses of yellowish green color, which in the sunlight fairly teem with bubbles of liberated oxygen. These dense masses of filaments furnish a home and shelter for a number of small animals, notably Haliplid beetle larvæ and punkie larvæ among insects; and entanglement by them is a peril to the lives of others, notably certain Mayfly larvæ (*Blasturus*). The rather large filaments afford a solid support for hosts of lesser sessile algæ; and their considerable accumulation of organic contents is preyed upon by many parasites. Their role is an important one in the economy of shoal waters, and its importance is due not alone to their power of rapid growth, but also to their staying qualities. They hold their own in all sorts of temporary waters by developing protected reproductive cells known as *zygospores*, which are able to endure temporary drouth, or other untoward conditions. Zygospores are formed by the fusion of the contents of two similar cells (the process is known as conjugation, whence the group name) and the development of a protective wall about the resulting reproductive body.   This rests for a time like a seed, and on germinating, produces a new filament by the ordinary process of cell division.   These filamentous forms share this reproductive process with the desmids, and despite the differences in external aspect it is a strong bond of affinity between the two groups.

The siphon algæ—This peculiar group of green algæ contains a few forms of little economic consequence but of great botanical interest.   The plant body grows out in long irregularly branching filaments which, though containing many nuclei, lack cross partitions.   The filaments thus resemble long open tubes, whence the name

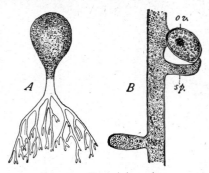

FIG. 42.   Two siphon algæ.

A, Botrydium; B, a small fruiting portion of a filament of *Vaucheria.; ov,* ovary; *sp,* spermary.

siphon algæ.   There are two common genera *Vaucheria* and *Botrydium* (fig. 42).   Both are mud-loving, and are found partly out of the water about as often as wholly immersed.   Vaucheria develops long, crooked, extensively interlaced filaments which occur in dense mats that have suggested the name "green felt." These felted masses are found floating in ponds, or lying on wet soil wherever there is light and a constantly moist atmosphere (as, for example, in greenhouses, where commonly found on the soil in pots). Botrydium is very different and much smaller.   It has an oval body with root-like branches growing out from the lower end to penetrate the mud.   It grows on the bottom in shoal waters, and remains exposed on the

mud after the water has receded, dotting the surface
thickly, as with greenish beads of dew.

The water net and its allies—The water net (*Hydrodictyon*) wherever found, is sure to attract attention by
its curious form. It is a cylindric sheet of lace-like
tissue, composed of slender green cells that meet at

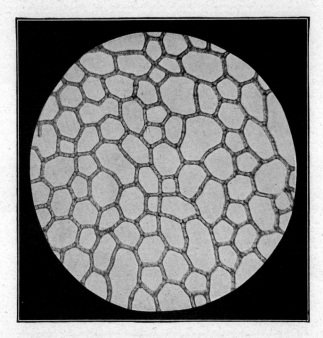

FIG. 43.   A rather irregular portion of a sheet of water net
(*Hydrodictyon*)

their ends, usually by threes, forming hexagonal meshes
like bobbinet (fig. 43). Such colonies may be as broad
as one's hand, or microscopic, or of any intermediate
size; for curiously enough, cell division and cell
growth are segregated in time. New colonies are
formed by repeated division of the contents of single

cells of the old colonies. A new complete miniature net is formed within a single cell; and after its escape from the old cell wall, it grows, not by further division, but by increase in size of its constituent cells.

Water net is rather local and sporadic in occurrence, but it sometimes develops in quantities sufficient to fill the waters of pools and small ponds.

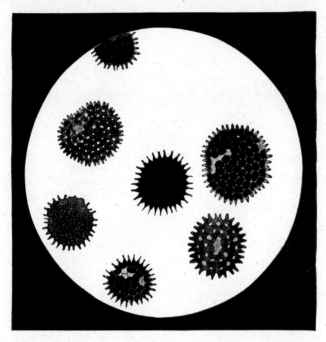

FIG. 44. *Pediastrum:* Several species from the plancton of Cayuga Lake.

Pediastrum is a closely related genus containing a number of beautiful species, some of which are common and widespread. The cells of a Pediastrum colony are arranged in a roundish flat disc, and those of the outermost row are usually prolonged into radiating points. Several species are shown in figure 44. In the open-meshed species the inner cells can be seen to meet by threes about the openings, quite as in the water net; but the cells are less elongate and the openings smaller. Five of the seven specimens shown in the figure lack these openings altogether.

New colonies are formed within single cells, as in Hydrodictyon. In our figure certain specimens show marginal cells containing developing colonies. One shows an empty cell wall from whence a new colony has escaped.

Other green algæ— We have now mentioned a few of the more strongly marked groups of the green algæ. There are other forms, so numerous we may not even name them here, many of which are common and widely dispersed. We shall have space to mention only a few of the more important among them,

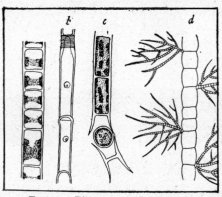

FIG. 45. Filamentous Green Algæ.

a, Ulothrix; b, Œdogonium, showing characteristic annulate appearance at upper end of cell; *c, Conferva (Tribonema); d, Draparnaldia.* (After West).

and we trust that the accompanying figures will aid in their recognition. Numerous and varied as they are, we will dismiss them from further consideration under a few arbitrary form types.

1. *Simple filamentous forms.* Of such sort are *Ulothrix, Œdogonium, Conferva,* etc., (fig. 45). Ulothrix is common in sunny rivulets and pools, especially in early spring, where its slender filaments form masses

half floating in the water. The cells are short, often no longer than wide, and each contains a single sheet of

Fig. 46. A spray of Cladophora, as it appears when outspread in the water, slightly magnified.

chlorophyl, lining nearly all of its lateral wall. *Œdogonium* is a form with stouter filaments composed of much longer cells, within which the chlorophyl is disposed in anastomosing bands. The thick cell walls, some of which show a peculiar cross striation near one end of the cell, are ready means of recognition of the members of this great genus. The filaments are attached when young, but break away and float freely in masses in quiet waters when older; it is thus they are usually seen. *Conferva* (Tribonema) abounds in shallow pools, especially in spring time. Its filaments are composed of elongate cells containing a number of separate disc-like chlorophyl bodies. The cell wall is thicker toward the ends of the cell, and the filaments tend to break across the middle, forming pieces (halves of two adjacent cells) which appear distinctly H-shaped in optic section. This is a useful mark for their recognition. It will be observed that these then are similar in form and habits to the filamentous conjugates discussed above, but they have not the peculiar form of chlorophyl bodies characteristic of that group. Œodgonium is remarkable for its mode of reproduction.

FIG. 47. Two species of Chætophora, represented by several small hemispherical colonies of *C. pisiformis* and one large branching colony of *C. incrassata.*

2. *Branching filamentous forms*—Of such sort are a number of tufted sessile algæ of great importance: *Cladophora*, which luxuriates in the dashing waterfall, which clothes every wave-swept boulder and pier with delicate fringes of green, which lays prostrate its pliant sprays (fig. 46) before each on-rushing wave, and lifts

them again uninjured, after the force of the flood is spent. And *Chætophora* (fig. 47; also fig. 89 on p. 182); which is always deeply buried under a transparent mass

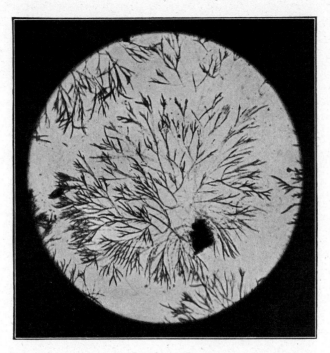

Fig. 48. Chætophora (either species) crushed and outspread in its own gelatinous covering and magnified to show the form of the filaments.

of gelatin; which forms little hemispherical hillocks of filaments in some species, and in one, extends outward in long picturesque sprays, but which has in all much the same form of plant body (fig. 48)—a close-set branching filament, with the tips of some of the branches ending in a long hyaline bristle-like point. Chætophora grows very abundantly in stagnant pools and ponds in midsummer, adhering to every solid support that offers, and it is an important part of the summer food of many of the lesser herbivores in such waters.

Then we must not omit to mention two that, if less important, are certainly no less interesting: *Draparnaldia* (fig. 45*d*) which lets its exceedingly delicate sprays trail like tresses among the submerged stones in spring-

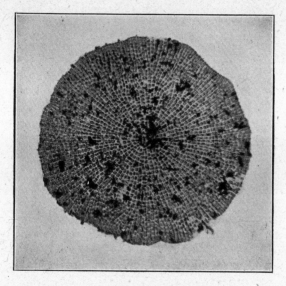

FIG. 49.   *Coleochæte scutata.*   "Green doily."

fed rivulets; and *Coleochæte* (fig. 49), which spreads its flattened branches out in one plane, joined by their edges, forming a disc, that is oftenest found attached to the vertical stem of some reed or bulrush.

Miscellaneous lesser green algæ—Among other green algæ, which are very numerous, we have space here for a mere mention of a few of the forms most likely to be met with, especially by one using a plancton net in open waters. These will also illustrate something of the remarkable diversity of form and of cell grouping among the lesser green algæ.

Botryococcus grows in free floating single or compound clusters of little globose green cells, held together in a scanty gelatinous investment. The clusters are sufficiently grape-like to have suggested the scientific name. They contain, when grown, usually 16 or 32 cells each. They are found in the open waters of bog pools, lakes,

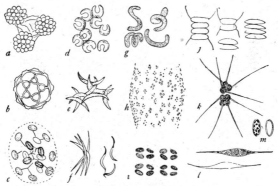

FIG. 50.   Miscellaneous green algæ (mostly after West).

a, Botryococcus; b, Cœlastrum; c, Dictosphœrium; d, Kirchnerella; e, Selenastrum; f, Ankistrodesmus falcatus; g, Ophiocytium; h, Tetraspora; i, Crucigenia; j, Scenedesmus;, k, Rhicteriella; l, Ankistrodesmus setigerus; m, Oocystis.

and streams, during the warmer part of the season, being most abundant during the hot days of August. When over-abundant the cells sometimes become filled with a brick-red oil. They occur sparingly in water-bloom.

Dictyosphœrum likewise grows in more or less spherical colonies of globose cells. The cells are connected together by dichotomously branching threads and all are enveloped in a thin spherical mass of mucus. The colonies are free floating and are taken in the plancton of ponds and lakes and often occur in the water-bloom.

Cœlastrum is another midsummer plancton alga that forms spherical colonies of from 8 to 32 cells; it has much firmer and thicker cell walls, and the cells are often angulate or polyhedral. New colonies are formed within the walls of each of the cells of the parent colony, and when well grown these escape by rupture or dissolution of the old cell wall. Our figure shows merely the outline of the cell walls of a 16-celled colony, in a species having angulate cells, between which are open interspaces. Kofoid found Cœlastrum occurring in a maximum of 10,800,000 per cubic meter of water in the Illinois River in August.

Crucigenia is an allied form having ovoid or globose cells arranged in a flat plate held together by a thin mucilaginous envelope. The cells are grouped in fours, but 8, 16, 32, 64 or even more may, when undisturbed, remain together in a single flat colony. During the warmer part of the season, they are common constituents of the fresh-water plancton, the maximum heat of midsummer apparently being most favorable to their development.

Scenedesmus is a very hardy, minute, green alga of wide distribution. There is hardly any alga that appears more commonly in jars of water left standing about the laboratory. When the sides of the jar begin to show a film of light yellowish-green, Scenedesmus may be looked for. The cells are more or less spindle-shaped, sharply pointed, or even bristle-tipped at the ends. They are arranged side by side in loose flat rafts of 2, 4 or 8 (oftenest, when not broken asunder, of 4) cells. They are common in plancton generally, especially in the plancton of stagnant water and in that of polluted streams, and although present at all seasons, they are far more abundant in mid and late summer.

Kirchnerella is a loose aggregate of a few blunt-pointed U-shaped cells, enveloped in a thick spherical mass of jelly. It is met with commonly in the plancton of larger lakes. *Selenastrum* grows in nearly naked clusters of more crescentic, more pointed cells which are found amid shore vegetation. *Ankistrodesmus* is a related, more slender, less crescentic form of more extensive littoral distribution. The slenderest forms of this genus are free floating, and some of them like *A. setigera* (fig. 50 *l*) are met with only in the plancton.

Richteriella is another plancton alga found in free floating masses of a few loosely aggregated cells. The cells are globose and each bears a few long bristles upon its outer face. Kofoid found Richteriella attaining a maximum of 36,000,000 per cubic meter of water in September, while disappearing entirely at temperatures below 60° F.

Oocystis grows amid shore vegetation, or the lighter species, in plancton in open water. The ellipsoid cells exist singly, or a few are loosely associated together in a clump of mucus. The cells possess a firm smooth wall which commonly shows a nodular thickening at each pole.

Ophiocytium is a curious form with spirally coiled multinucleate cells. The bluntly rounded ends of the cells are sometimes spine-tipped. These cells sometimes float free, sometimes are attached singly, sometimes in colonies. Kofoid found them of variable occurrence in the Illinois River, where the maximum number noted was 57,000,000 per cubic meter occurring in September. The optimum temperature, as attested by the numbers developing, appeared to be about 60° F.

Tetraspora—We will conclude this list of miscellanies with citing one that grows in thick convoluted strings and loose ropy masses of gelatin of considerable size. These masses are often large enough to be recognized with the unaided eye as they lie outspread or hang down upon trash on the shores of shoal and stagnant waters. Within the gelatin are minute spherical bright green cells, scattered or arranged in groups of fours.

BLUE-GREEN ALGÆ (*Cyanophyceæ* or *Myxophyceæ*). The "blue-greens" are mainly freshwater algæ, of simple forms. The cells exist singly, or embedded together in loose gelatinous envelope or adhere in flat rafts or in filaments. Their chlorophyl is rather uniformly distributed over the outer part of the cell (quite lacking the restriction to specialized chloroplasts seen in the true green-algæ) and its color is much modified by the presence of pigment (*phycocyanin*), which gives to the cell usually a pronounced bluish-green, sometimes, a reddish color.

Blue-green algæ exist wherever there is even a little transient moisture—on tree trunks, on the soil, in lichens, etc.; and in all fresh water they play an important role, for they are fitted to all sorts of aquatic situations, and they are possessed of enormous reproductive capacity. Among the most abundant plants in the water world are the *Anabænas* (fig. 179), and other blue-greens that multiply and fill the waters of our lakes in midsummer, and break in "water-bloom" covering the entire surface and drifting with high winds in windrows on shore. Such forms by their decay often give to the water of reservoirs disagreeable odors and bad flavors, and so they are counted noxious to water supplies.

There are many common blue-greens, and here we have space to mention but a few of the more common forms. Two of the loosely colonial forms composed of spherical cells held together in masses of mucus are *Cælosphærium* and *Microcystis*. Both these are often associated with Anabæna in the water-bloom. Cœlosphærium is a spherical hollow colony of microscopic size. It is a loose association of cells, any of which on separation is capable of dividing and producing a new colony. Microcystis (fig. 51*A*) is a mass of smaller cells, a very loose colony that is at first more or less spherical but later becomes irregularly lobed and branching. Such old colonies are often large enough to be observed with the naked eye. They are found most commonly in late summer, being hot weather forms. When abundant these two are often tossed by the waves upon rocks along the water's edge, and from them the dirty blue-green deposit that is popularly known as "green paint."

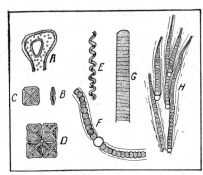

FIG. 51. Miscellaneous blue-green algæ (mostly after West).

A, Microcystis (Clathrocystis); *B, C, D, Tetrapedia; E, Spirulina; F, Nostoc; G, Oscillatoria; H, Rivularia.*

Among the members of this group most commonly seen are the motile blue-greens of the genus *Oscillatoria* (fig. 51*G*). These grow in dense, strongly colored tufts and patches of exceedingly slender filaments attached to the bottoms and sides of watering troughs, ditches and pools, and on the beds of ponds however stagnant. They thickly cover patches of the black mud bottom and the formation of gases beneath them disrupts their attachment and the broken flakes of bottom slime that they hold together, rise to the surface and float there, much to the hurt of the appearance of the water.

The filaments of Oscillatoria and of a few of its near allies perform curious oscillating and gliding movements. Detached filaments float freely in the open water, and during the warmer portion of the year, are among the commoner constituents of the plancton.

There are a number of filamentous blue-greens that are more permanently sessile, and whose colonies of filaments assume more definite form. *Rivularia* is typical of these. *Rivularia* grows in hemispherical gelatinous lumps, attached to the leaves and stems of submerged seed plants. In autumn it often fairly smothers the beds of hornwort (*Ceratophyllum*) and water fern (*Marsilea*) in rich shoals. Rivularia is

FIG. 52. Colonies of Rivularia on a disintegrating Typha leaf.

brownish in color, appearing dirty yellowish under the microscope. Its tapering filaments are closely massed together in the center of the rather solid gelatinous lump. The differentiation of cells in the single filament is shown in fig. 51*H*. Such filaments are placed side by side, their basal heterocysts close together, their tips diverging. As the mass grows to a size larger than a pea it becomes softer in consistency, more loosely attached to its support and hollow. Strikingly different in form and habits is the raftlike *Merismopædia* (fig. 53). It is a flat colony of shining blue-green cells that divide in two planes at right angles to each other, with striking regularity. These rafts of cells drift about freely in open water, and are often taken in the plancton, though rarely in great abundance. They settle betimes on the leaves of the larger water plants, and may be discovered with a pocket lens by searching the sediment shaken therefrom.

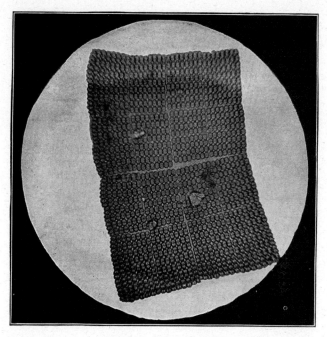

FIG. 53. Merismopædia.

RED and BROWN ALGÆ (*Rhodophyceæ* and *Phæophyceæ*) —These groups are almost exclusively marine. A few scattering forms that grow in fresh water are shown in figure 54. *Lemanea* is a torrent-inhabiting form that grows in blackish green tufts of slender filaments, attached to the rocks in deep clear mountain streams where the force of the water is greatest. It is easily recognizable by the swollen or nodulose appearance of the ultimate (fruiting) branches. *Chantransia* is a beautiful purplish-brown, extensively branching form that is more widely distributed.   It is common in clear flowing streams.   It much resembles Cladophora in manner of growth but is at once distinguished by its color.

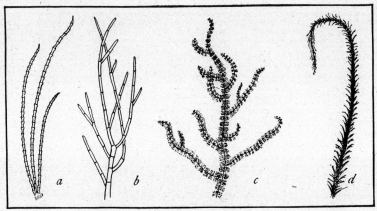

FIG. 54.   Red and brown algæ (after West).

a, Lemanea;   b, Chantransia;   c, Batrachospermum;   d, Hydrurus.

Batrachospermum is a freshwater form of wide distribution, with a preference for spring brooks, though occurring in any water that is not stagnant.   It grows in branching filaments often several inches long, enveloped in a thick coat of soft transparent mucus.   The color is bluish or yellowish-green, dirty yellow or brownish. Attached to some stick or stone in a rivulet its sprays, of more than frond-like delicacy, float freely in the water.

Hydrurus grows in branched colonies embedded in a tough mucilage, attached to rocks in cold mountain streams.   The colonies are often several inches long. Their color is olive green.   They have a plumose appearance, and are of very graceful outline.

The stoneworts (*Characeæ*).—This group is well represented in freshwater by two common genera, well known
to every biological laboratory student, *Chara* and
Nitella. Both grow in protected shoals, and in the
borders of clear lakes at depths below the heavy beating
of the waves. Both are brittle and cannot withstand

FIG. 55. Nitella glomerulifera.

wave action. Both prefer the waters that flow off from
calcareous soils, and are oftenest found attached to a
stony bottom.

The stoneworts, are the most specialized of the fresh-
water algæ: indeed, they are not ranked as algæ by
some botanists. In form they have more likeness to
certain land plants than to any of the other algæ.

They grow attached to the soil.   They grow to considerable size, often a foot or more in length of stem.   They grow by apical buds, and they send out branches in regular whorls, which branch and branch again, giving the plant as a whole a bushy form.   The perfect regularity of the whorled branches and the brilliant coloration of the little spermaries borne thereon, doubtless have suggested the German name for them of "Candelabra plants."

The stoneworts are so unique in structure and in reproductive parts that they are easily distinguished from other plants.   The stems are made up of nodes and internodes.   The nodes are made up of short cells from which the branches arise.   The internodes are made up of long cells (sometimes an inch or more long), the central one of which reaches from one node to another.   In Nitella there is a single naked internodal cell composing entirely that portion of the stem.   In Chara this axial cell is covered externally by a single layer of slenderer cortical cells wound spirally about the central one.   A glance with a pocket lens will determine whether there is a cortical layer covering the axial internodal cell, and so will distinguish Chara from Nitella.   Chara is usually much more heavily incrusted with lime in our commoner species, and in one very common one, *Chara fœtida*, exhales a bad odor of sulphurous compounds.

The sex organs are borne at the bases of branchlets. There is a single egg in each ovary, charged with a rich store of food products, and covered by a spirally wound cortical layer of protecting cells.   These, when the egg is fertilized form a hard shell which, like the coats of a seed, resist unfavorable influences for a long time. This fruit ripens and falls from the stem.   It drifts about over the bottom, and later it germinates.

At the apex of the ovary is a little crown of cells, between which lies the passageway for the entrance of the sperm cell at the time of fertilization. This crown is composed of five cells in Chara; of ten cells in Nitella. It is deciduous in Chara; it is persistent in Nitella.

The stoneworts, unlike many other algæ, are wonderfully constant in their localities and distribution, and regular in their season of fruiting. They cover the same hard bottoms with the same sort of gray-green meadows, year after year, and although little eaten by aquatic animals, they contribute important shelter for them, and they furnish admirable support for many lesser epiphytes.

CHLOROPHYLLESS WATER PLANTS, BACTERIA AND FUNGI

Nature's great agencies for the dissolution of dead organic materials, in water as on land, are the plants that lack chlorophyl. They mostly reproduce by means of spores that are excessively minute and abundant, and that are distributed by wind or water everywhere; consequently they are the most ubiquitous of organisms. They consume oxygen and give off carbon dioxide as do the animals, and having no means of obtaining carbon from the air, must get it from carbonaceous organic products—usually from some carbohydrate, like sugar, starch, or cellulose. Some of them can utilize the nitrogen supply of the atmosphere but most of them must get nitrogen also from the decomposition products of pre-existing proteins. Many of them produce active ferments, which expedite enormously the dissolution of the bodies of dead plants and animals. Some bacteria live without free oxygen.

It follows from the nature of their foods, that we find these chlorophylless plants abounding where there is the best supply of organic food stuffs: stagnant pools filled with organic remains, and sewers laden with the city's waste. But there is no natural water free from them.   Let a dead fly fall upon the surface of a tumbler of pond water and remain there for a day or two and it becomes white with water mold, whose spores were present in the water.   Let any organic solution stand exposed and quickly the evidence of rapid decomposition appears in it.   Even the dilute solutions contained in a laboratory aquarium, holding no organic material other than a few dead leaves will often times acquire a faint purple or roseate hue as chromogenic bacteria multiply in them.

Bacteria—A handful of hay in water will in a few hours make an infusion, on the surface of which a film of "bacterial jelly" will gather.   If a bit of this "jelly" be mounted for the microscope, the bacteria that secrete it may be found immersed in it, and other bacteria will be found adherent to it.   All the common form-types, *bacillus, coccus* and *spirillum* are likely to be seen readily.   Thus easy is it to encourage a rich growth of water bacteria.   Among the bacteria of the water are numerous species that remain there constantly (often called "natural water bacteria"), commingled at certain times and places with other bacteria washed in from the surface of the soil, or poured in with sewage.   From the last named source come the species injurious to human health.   These survive in the open water for but a short time.   The natural water bacteria are mainly beneficial; they assist in keeping the world's food supply in circulation.   Certain of them begin the work of altering the complex organic substances.   They attack the proteins and produce from them ammonia and various ammoniacal compounds.   Then other bacteria, the so-called "nitrifying" bacteria attack the ammonia, changing it to simpler compounds.   Two kinds of bacteria successively participate in this: one kind oxidizes the ammonia to nitrites; a second kind oxidizes the nitrites to nitrates. By these successive operations the stores of nitrogen that are gathered together within the living bodies of plants and animals are again released for further use. The simple nitrates are proper food for the green algæ, with whose growth the cycle begins again. And those bacteria which promote the processes of putrefaction, are thus the world's chief agencies for maintaining undiminished growth in perpetual succession.

Bacteria are among the smallest of organisms. Little of bodily structure is discoverable in them even with high powers of the microscope, and consequently they are studied almost entirely in specially prepared cultures, made by methods that require the technical training of the bacteriological laboratory for their mastery. Any one can find bacteria in the water, but only a trained specialist can tell what sort of bacteria he has found; whether pathogenic species like the typhoid bacillus, or the cholera spirillum; or whether harmless species, normal to pure water.

The higher bacteria—Allied to those bacilli that grow in filaments are some forms of larger growth, known as *Trichobacteria*, whose filaments sometimes grow attached in colonies, and in some are free and motile. A few of those that are of interest and importance in fresh-water will be briefly mentioned and illustrated here.

*Leptothrix** (Fig. 56*a*, *b* and *c*) grows in tufts of slender, hairlike filaments composed of cylindric cells surrounded by a thin gelatinous sheath. In reproduction the cells are transformed directly into spores (*gonidia*) which escape from the end of the sheath and, finding favoring conditions, grow up into new filaments.

---

*Known also as *Streptothrix* and *Chlamydothrix*.

Crenothrix (Fig. 56 *d*, *e* and *f*) is a similar unbranched sessile form which is distinguished by a widening of the filaments toward the free end. This is caused by a division of the cells in two or three planes within the sheath of the filament, previous to spore formation. Often by the germination of spores that have settled upon the outside of the old sheaths and growth of new filaments therefrom compound masses of appreciable

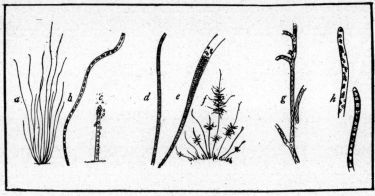

FIG. 56.   Trichobacteria.

a, *b*, *c*, *Leptothrix* (*Streptothrix*, or *Chlamydothrix*). *a*, a colony; *b*, a single filament; *c*, spore formation; *d*, *e*, *f*, *Crenothrix*; *d*, a single growing filament; *e*, a fruiting filament; *f*, a compound colony; *g*, *Cladothrix*, a branching filament; *h*, *Beggiatoa*, younger and older filaments, the latter showing sulphur granules, and no septa between cells of the filament.

size are produced. In the sheaths of the filaments a hydroxide of iron is deposited (for Crenothrix possesses the power of oxidizing certain forms of iron); and with continued growth the deposits sometimes become sufficient to make trouble in city water supply systems by stoppage of the pipes. In nature, also, certain deposits of iron are due to this and allied forms properly known as iron bacteria. *Cladothrix* (Fig. 56 *g*), is a related form that exhibits a peculiar type of branching in its slender cylindric filaments.

Beggiatoa (fig. 56 *h*) is the commonest of the so-called sulfur bacteria. Its cylindric unbranched and unattached filaments are motile, and rotate on the long axis with swinging of the free ends. The boundaries between the short cylindric cells are often obscure, especially when (as is often the case) the cells are filled with highly refractive granules of sulfur. Considerable deposits of sulfur, especially about springs, are due to the activities of this and allied forms.

Water molds—True fungi of a larger growth abound in all fresh waters, feeding on almost every sort of organic substance contained therein. The commonest of the water molds are the Saprolegnias, that so quickly overgrow any bit of dead animal tissue which may chance to fall upon the surface of the water and float there. If it be a fly, in a day or two its body is surrounded by a white fringe of radiating fungus filaments, outgrowing from the body. The tips of many of these filaments terminate in cylindric sporangia, which when

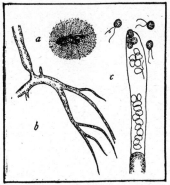

FIG. 57.   A common water mold, *Saprolegnia.* (After Engler and Prantl.)

a, a colony growing on a dead fly; *b*, a bit of the mycelium that penetrates the fly's body; *c*. a fruiting tip, with escaping swarm spores.

mature, liberate from their ruptured tips innumerable biciliated free-swimming swarm spores. These wander in search of new floating carcases, or other suitable food. Certain of these water molds attack living fishes, entering their skin wherever there is a a slight abrasion of the surface, and rapidly producing diseased conditions. These are among the worst pests with which the fish culturist has to contend. They attack also the eggs of fishes during their incubation, as shown in a figure in a later chapter.

Most water molds live upon other plants. Even the Saprolegnias have their own lesser mold parasites. Many living algæ, even the lesser forms like desmids and diatoms are subject to their attack. Fine cultures of such algæ are sometimes run through with an epidemic of mould parasites and ruined.

THE HIGHER PLANTS

(*Mossworts, Fernworts and Seed Plants*)

In striking contrast with the algæ, the higher plants live mainly on land, and the aquatics among them are restricted in distribution to shoal waters and to the vicinity of shores. There is much in the bodily organization of nearly all of them that indicates ancestral adaptation to life on land. They have more of hard parts, more of localized feeding organs, more of epidermal specialization, and more differentiation of parts in the body, than life in the water demands.

FIG. 58. The marsh mallow, *Hibiscus Moscheutos.*

They occupy merely the margins of the water. A few highly specialized genera, well equipped for withstanding partial or complete submergence occupy the shoals and these are backed on the shore line by a mingled lot of semi-aquatics that are for the most part but stray members of groups that abound on land. Often they are single members of large groups and are sufficiently distinguished from their fellows by a name indicating the kind of wet place in which they grow. Thus we know familiarly the floating riccia, the bog mosses, the brook speedwell, the water fern and water cress, the marsh bell flower and the marsh fern, the swamp horsetail and the swamp iris, etc. All these and many others are stragglers from large dry land groups. That readaptation to aquatic life has occurred many times independently is indicated by the fact that the more truly aquatic families are small and highly specialized, and are widely separated systematically.

Bryophytes—Both liverworts and mosses are found in our inland waters, though the former are but sparingly represented. Two simple Riccias, half an inch long when grown, are the liverworts most commonly found. One, *Riccia fluitans*, grows in loose clusters of flat slender forking sprays that drift about so freely that fragments are often taken in pond and river plancton. The larger unbroken more or less spherical masses of sprays are found rolling with the waves upon the shores of muddy ponds. The other, *Ricciocarpus natans*, has larger and thicker sprays of green and purple hue, that float singly upon the surface, or gather in floating masses covering considerable areas of quiet water. They are not uncommonly found in springtime about the edges of muddy ponds. Underneath the flat plant body there is a dense brush of flattened scales.

Water mosses are more important. The most remarkable of these are the bog mosses (*Sphagnum*). These cover large areas of the earth's surface, especially in northern regions, where they chiefly compose the thick soft carpet of vegetation that overspreads open bogs and coniferous swamps. They are of a light grey-green color, often red or pink at the tips. These mosses do not grow submerged, but they hold immense quantities of water in their reservoir cells, and are able to absorb water readily from a moist atmosphere; so they are always wet. Supported on a framework of entangled rootstocks of other higher plants, the bog mosses extend out over the edges of ponds in floating mats, which sink under one's weight beneath the water level and rise again when the weight is removed. The part of the mat which the sphagnum composes consists of erect, closely-placed, unbranched stems, like those shown in fig. 59, which grow ever upward at their tips,

Fig. 59. Bog moss, *Sphagnum.*

and die at the lower ends, contributing their remains to the formation of beds of peat.

The leaves of Sphagnum are composed of a single layer of cells that are of two very different sorts. There are numerous ordinary narrow chlorophyl-bearing cells, and, lying between these, there are larger perforate reservoir cells, for holding water.

The true water mosses of the genus *Fontinalis* are fine aquatic bryophytes. These are easily recognized, being very dark in color and very slender. They grow in spring brooks and in clear streams, and are often seen in great dark masses trailing their wiry stems where the current rushes between great boulders or leaps into foam-flecked pools in mountain brooks.

Another slender brook-inhabiting moss is *Fissidens julianum*, which somewhat resembles Fontinalis, but which is at once distinguished by the deeply channeled bases of its leaves, which enfold the stem. The leaves are two ranked and alternate along the very slender flexuous stem, and appear to be set with edges toward it.

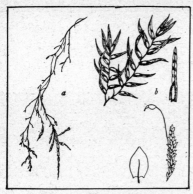

FIG. 60.   Water mosses.

a, Fontinalis; b, Fissidens julianum, with a single detached leaf, more enlarged; *c, Rhynchostegium rusciforme*, with a single detached leaf at the left.   (After Grout.)

There are also a few lesser water mosses allied to the familiar trailing hypnums, so common in deep woods. They grow on stones in the bed of brooks. They cover the face of the ledges over which the water pours in floods and trickles in times of drouth, as with a fine feathery carpet of verdure that adds much to the beauty of the little waterfalls. They give shelter in such places to an interesting population of amphibious animals, as will be noted in chapter VI, following. The leaves of the hypnums are rather short and broad, and in color they are often very dark—often almost black.*

*Grout has given a few hints for the recognition of these "Water-loving hypnums" in his *Mosses with a Hand Lens*, 2d edition, p. 128. New York, 1905.

There are also a few hypnums found intermixed with sphagnum on the surface of bogs, and as everyone knows there are hosts of mosses in all moist places in woods and by watersides.

FIG. 61.   Two floating leaves of the "water shamrock," *Marsilea*, in the midst of a surface layer of duck-meat (*Spirodela polyrhiza*).   "Lemna" on fig. 62.

Pteridophytes—Aquatic fernworts are few and of very unusual types.   There is at least one of them, however, that is locally dominant in our flora.   *Marsilea*, the so-called water shamrock or water fern, abounds on the sunny shoals of muddy bayous about Ithaca and in many places in New England. It covers the zone between high and low water, creeping extensively over the banks that are mostly exposed, and there forming a most beautiful ground cover, while producing longer leaf-stalks where submerged. These leaf-stalks carry the beautiful four-parted leaf-blades to the surface where they float gracefully. Fruiting bodies the size

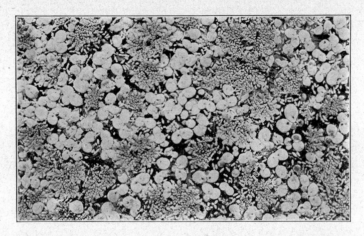

FIG. 62. Floating plants: The largest branching colonies are *Azolla;* the smallest plants are *Wolffia;* those of intermediate size are *Lemna minor.* Photo by Dr. Emmeline Moore.

of peas are produced in clusters on the creeping stems above the water line, often in very great abundance.

Then there are two floating pteridophytes of much interest. *Salvinia,* introduced from Europe, is found locally along our northeastern coast, and in the waters of our rich interior bottom lands the brilliant little *Azolla* flourishes. Azolla floats in sheltered bogs and back waters, intermingled with duckweeds. It is reddish in color oftener than green and grows in minute mosslike pinnately branched sprays, covered with closely overlapping two-lobed leaves, and emits a few rootlets from the under side which hang free in the water. In the back waters about the Illinois Station at Havana, Illinois, Azolla forms floating masses often several feet in diameter, of bright red rosettes.

Shoreward there are numerous pteridophytes growing as rooted and emergent aquatics; the almost grasslike *Isoetes*, and the marsh horsetails and ferns, but these latter differ little from their near relatives that live on land.

Aquatic Seed Plants—These are manifestly land plants in origin. They have much stiffening in their stems. They have a highly developed epidermal system, often retaining stomates, although these can be no longer of service for intake of air. They effect fertilization by means of sperm nuclei and pollen tubes, and not by free swimming sperm cells.

Seed plants crowd the shore line, but they rapidly diminish in numbers in deepening water. They grow thickest by the waterside because of the abundance of air moisture and light there available. But too much moisture excludes the air and fewer of them are able to grow where the soil is always saturated. Still fewer grow in standing water and only a very few can grow wholly submerged. Moreover, it is only in protected shoals that aquatic seed plants flourish. They cannot withstand the beating of the waves on exposed shores. Their bodies are too highly organized, with too great differentiation of parts. Hence the vast expanses of open waters are left in possession of the more simply organized algæ.

An examination of any local flora, such as that of the Cayuga Lake Basin* will reveal at once how small a part of the population is adapted for living in water.

*The following data are largely drawn from Dudley's *Cayuga Flora*, 1886.

In this area there are recorded as growing without cultivation 1278 species. Of these 392 grow in the water. However, fewer than forty species grow wholly submerged, with ten or a dozen additional submerged except for floating leaves. Hardly more than an eighth, therefore, of the so-called "aquatics" are truly aquatic in mode of life: the remaining seven-eighths grow on shores and in springs, in swamps and bogs, in ditches, pools, etc., where only their roots are constantly wet.

The aquatic seed plants are representative of a few small and scattered families. Indeed, the only genus having any considerable number of truly aquatic species is the naiad genus Potamogeton. Other genera of river-weeds, or true pond weeds, are small, scattered and highly diversified. They bear many earmarks of

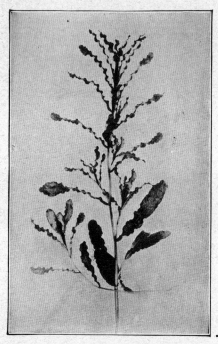

Fig. 63. The ruffled pond-weed; *Potamogeton crispus*, one of the most ornamental of fresh water plants.

independent adaptation to the special situations in the water which they severally occupy. In the economy of nature the Potamogetons or river weeds constitute the most important single group of submerged seed plants. They are rooted to the bottom in most shoal waters, and compose the greater part of the larger water meadows within our flora. They have alternate leaves and slender flexuous stems that are often incrusted with lime.

There are evergreen species among the Potamogetons, and other species that die down in late summer. There are broad leaved and narrow leaved species. There are a few, like the familiar *Potamogeton natans* whose uppermost leaves float flat upon the surface, but the more important members of the genus live wholly submerged. Tho seed-plants, they mainly reproduce vegetatively, by specialized reproductive buds that are developed in the growing season, and are equipped with stored starch and other food reserves, fitting them when detached for rapid growth in new situations. These reproductive parts are developed in some species as tuberous thickenings of underground parts; in others as burr-like clusters of thickened apical buds; and in still others they are mere thickenings of detachable twigs.

The Potamogetons enter largely into the diet of wild ducks and aquatic rodents and other lesser aquatic herbivores. They are as important for forage in the water as grasses are on land.

Other naiads are *Nais* (fig. 85) and *Zannichellia*.

Eel-grass (*Vallisneria*) is commonly mixed with the pond weeds in lake borders and water meadows. Eel-grass is apparently stemless and has long, flat, flexuous, translucent, ribbonlike leaves, by which it is easily recognized. The duckweeds (Lemnaceæ, figs. 61 and 62) are peculiar free-floating forms in which the plant body is a small flat thallus, that drifts about freely on the surface in sheltered coves, mingled with such liverworts as Ricciocarpus, with such fernworts as Azolla, with seeds, eel-grass flowers, and other flotsam. There are definite upper and lower surfaces to the thallus with pendant roots beneath hanging free in the water. Increase is by budding and outgrowth of new lobes from pre-existing thalli. Flowering and seed production are of rare occurrence.

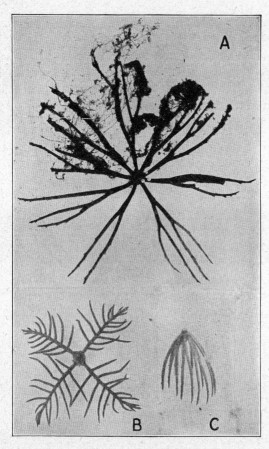

FIG. 64. Leaf-whorls.

A, and *C,* the hornwort (*Ceratophyllum*); *B,* the water milfoil (*Myriophyllum*). *A* is an old leaf, the upper half normally covered with algæ and silt; the lower half cleaned, save for a closely adherent dwelling-tube of a midge larva in the fork at the right. *C,* is a young partly expanded leaf whorl from the apical bud.

The water lily family includes the more conspicuous of the broad-leaved aquatics, which pre-empt the rich bottom mud with stout root stocks, and heavily shade the water with large shield-shaped leaves, either floating upon the surface, as in the water shield and water lilies or lifted somewhat above it, as in the spatterdock and the lotus. They are long-lived perennials, requiring a rich muck soil to root in. These are distinguished for the beauty and fragrance of their flowers.

The bladderworts (*Utricularia*) comprise another peculiar group. They are free-floating, submerged plants with long, flexuous branching stems that are thickly clothed with dissected leaves. Attached to the leaves are the curious traps or "bladders" (discussed in Chapt. VI) which have suggested the group name. Being unattached they frequent the still waters of sheltered bays and ponds where they form beautiful feathery masses of green. They shoot up stalks above the surface bearing curious bilabiate flowers.

Fig. 65. The water weed, *Philotria* (*Anacharis* or *Elodea*), with two young black-and-green-banded nymphs of the dragonfly *Anax* on its stem, and a snail, *Planorbis*, on a leaf.

The hornwort (*Ceratophyllum*) is another non-rooting water plant that grows wholly submerged and branching. It is coarser, however, and hardier than Utricularia and much more widespread. Its leaves are stiff, repeatedly forking, and spinous-tipped (fig. 64 *A* and *C*).

The water milfoils (*Myriophyllum*) are rooted aquatics, superficially similar to the hornwort but distinguishable at a glance by the simple pinnate branching of the softer leaves (fig. 64*B*).

Then there are a few very common aquatics that form patches covering the beds of lesser ponds, bogs and pools.   The common water weed, *Philotria*, (fig. 65), with its neat little leaves regularly arranged in whorls of threes; and two water crowfoots, *Ranunculus*, (fig. 66), white and yellow, with alternate finely dissected leaves; and the water purslane, *Ludvigia palustris*, with its closely-crowded opposite ovate leaves are found here.

These are the common plants of the waterbeds about Ithaca.   They are so few one may learn them quickly, for so strongly marked are they that a single spray or often a single leaf is adequate for recognition.

FIG. 66.   A leaf of the white water-crow-foot, *Ranunculus*.

Then there are three small families so finely adapted to withstanding root submersion that they dominate all our permanent shoals and marshes.   These are (1) the Typhaceæ including the cat-tails and the bur-reeds, which form vast stretches of nearly clear growth, as discussed in the last chapter; (2) the Alismaceæ, including arrow heads and water plantain, and (3) the Pontederiaceæ, represented by the beautiful blue pickerel-weed.   All these are shown in their native haunts in the figures of chapter VI.

Another family of restricted aquatic habitat is the Droseraceæ, the sun-dews, which grow in the borders of sphagnous upland bogs.   They are minute purplish-tinted plants whose leaves bear glandular hairs.

Few other families are represented in the water by more than a small proportion of their species.   Those families are best represented whose members live chiefly on low grounds and in moist soil. A few rushes (Juncaceæ) invade the water on wave-washed shores at fore front of the standing aquatics. A few sedges

FIG. 67. Fruit clusters of four emergent aquatic seed plants; arrow-arum (*Peltandra*), pickerel-weed (*Pontederia*), burr-reed (*Sparganium*), and sweet flag (*Acorus*).

(Carices) overrun flood-plains or fringe the borders of ditches. A very few grasses preëmpt the beds of shallow and impermanent pools. A few aroids, such as arrow arum and the calla adorn the boggy shores. A few heaths, such as, Cassandra and Andromeda overspread the surface of upland sphagnum bogs with dense

levels of shrubs, and numerous orchids occupy the surface of the bog beneath and between the shrubs.
Willows and alders fringe all the streams, associated
there with a host of representatives of other families
crowding down to the waterside. A few of these on
account of their usefulness or their beauty, we shall
have occasion to consider in a subsequent chapter.

Such are the dominant aquatic seed plants in the
Cayuga Basin; and very similar are they over the
greater part of the earth. The semi-aquatic representatives of the larger families are few and differ little
from their terrestrial relatives: the truly aquatic
families are small and highly diversified.

ANIMALS

MANY of the lower groups of
animals are wholly aqua-
tic, never having de-
parted from their ances-
tral abode. Other groups
are in part adapted to
life on land. A few
others, after becoming fit
for terrestrial life, have
been readapted in part to
life in the water. Aqua-
tic insects and mammals,
especially, give evidence
of descent from terres-
trial ancestors. As with
plants, so with animals, it is the lower groups that
are predominantly aquatic. The simplest of animals
are the protozoans; so with these we will begin.

Protozoans—One of the best known animals in the world, one that is pedagogically exploited in every biological laboratory, is the *Amœba* (fig. 69a). Plastic, ever changing in form and undifferentiated in parts, this is the animal that is the standard of comparison among things primitive. Its name has become a household word, and an every-day figure of speech. A little living one-celled mass of naked pro-toplasm, that creeps freely about amid the ooze of the pond bottom, and feeds on organic foods. It grows just large enough to be recog-nized by the naked eye when in most favorable light, as when creeping up the side of a culture jar: on the pond bottom it is undiscoverable and a microscope is essential to study it.

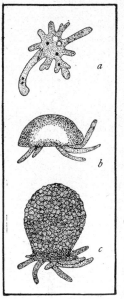

Related to Amœba are several common shell-bearing forms of the group of Sarcodina (Rhizopoda) that often become locally abun-dant. *Difflugia* (fig. 69c) forms a flask-shaped shell composed of mi-nute granules, that, magnified, look like grains of sand stuck together

FIG. 69. Protozoans.
a, Amœba; b, Arcella; c Difflugia.

over the outside. The soft amœba-like body protrudes in pseudopodia from the mouth of the flask, when travel-ing or foraging, or withdraws inside when disturbed. *Arcella* (fig. 69b) secretes a broadly domeshaped shell, having a concave bottom, in the center of which is the hole whence dangle the clumsy pseudopodia. One species of Arcella, shown in the following figure, has the margin of the shell strongly toothed. Both of these genera, and other shell-bearing forms, secrete

F𝐢ɢ. 70.   *Arcella dentata.*
Through the central opening
there is seen a diatom, recently swallowed.

bubbles of gas within their
shells whereby they are caused
to float.   Thus they are often
taken in the plancton net from
open water of the ponds and
streams.

Other protozoans that have
the body more or less covered with vibratile cilia (Cil-
iata), are very common in
fresh water, especially in ponds
and pools.   Best known of
these is *Paramecium*, (fig.
71a) another familiar biolog-
ical-laboratory "type" that
grows abundantly in plant infusions.   It is found in
stagnant pools, swimming near the surface.   There
are many species of Paramecium.   Some of them and
some members of allied genera are characteristic of
polluted waters.   Other allied genera are parasitic,
and live within the bodies of the
higher animals.   *Stentor* is (as the
name signifies) a more or less
trumpet-shaped ciliate protozoan,
that may detach itself and swim
freely about, but that is ordinarily attached by its slender
base to some support.   Its base
is in some species surrounded by
a soft gelatinous transparent
lorica, as shown in the figure.
Some species are of a greenish
color.   Stentor and Paramecium,
tho unicellular, are quite large
enough to be seen (as moving
specks) with the unaided eye.

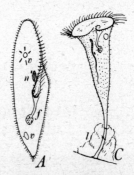

F𝐢ɢ. 71.   Ciliate pro-
tozoans.

A, *Paramœcium;*   *n,* nucleus;   *v, v,* vacuoles;   *f,*
food-ball at the bottom of
the rudimentary esopha-
gus;   *C, Stentor; l,* lorica.

Cothurnia (fig. 73c) is a curious double form that is often found attached to the stems of water weeds. The two cells of unequal height are surrounded by a thin transparent lorica. For beauty of form and delicacy or organization it would be hard to find anything surpassing this little creature.

Vorticella and its allies are among the commonest and most ubiquitous of protozoans. They are sessile and stalked, with some portion or all of the base contractile. Vorticella forms clusters of many separate individuals, while Epistylis forms branching, tree-like compound colonies (fig. 72). Oftentimes they completely clothe twigs and grass stems lying in the water, as with a white fringe. Often they cluster about the appendages of crustaceans and insects, or thickly clothe their shells. Sometimes they cling to floating algal filaments in the water-bloom (see fig. 179 on p. 295).

FIG. 72. A colony of *Epistylis.*

The dark object on the side of the stalk is an egg, probably the egg of a rotifer.

Ophrydium forms colonies of a very different sort. Numerous weak-stalked individuals have their bases imbedded in a roundish mass of gelatin. The colonies lie scattered about over the bottom of a lake or pond. They are roundish, or often rather shapeless masses varying in size from mere specks up to the dimensions of a hen's egg. In the summer of 1906 the marl-strewn shoals of Walnut Lake in Michigan were so thickly covered that a boat-load of the soft greenish-white colonies could easily have been gathered from a small area of the bottom.

Other forms of protozoa there are in endless variety. We cannot even name the common ones here: but we will mention two that are very different from the fore-

going in form and habit. *Podophrya* will often be encountered by searching the backs of aquatic insects or the
sides of submerged twigs, or other solid support, to which
it is attached.   It is sessile, and reaches out its suctorial
pseudopodia in search of soft-bodied organisms that are
its prey.

Anthophysa is a curious sessile form that is common
in polluted waters.   It forms very minute spherical
colonies that are attached to the transparent tip of a

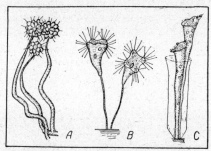

FIG. 73.   Three sessile protozoans.
A, Anthophysa; *B*, Podophrya; *C*, Cothurnia.

rather thick brownish stalk.   The stalk increases in
length and diameter with age, occasionally forking when
the colony divides.   It soon becomes much more conspicuous than the colonies it carries.   It often persists
after the animals are dead and gone.   After a vigorous
growth, the accumulated stalks sometimes cover every
solid support as with a soft flocculent brownish fringe.

Besides these and other free-living forms, there are
parasitic Protozoa whose spores get into the water.
Some of these are pathogenic; many of them have
changes of host; all of them are biologically interesting;
but we have not space for their consideration here.
We must content ourselves with the above brief
mention of a few of the more common and interesting
free-living forms.

METAZOANS

Hydras are the only common fresh-water representatives of the great group of Cœlenterates, so abundant in the seas; and of hydras there are but a few species. Two of these, the common green and brown ones, *H. virdis* and *H. fusca*, are well enough known, being among the staples of every biological laboratory. Pedagogically it is a matter of great good fortune that this little creature lives on, a common denizen of fresh-water pools; for its two-layered sac-like body represents well the simplest existing type of metazoan structure.

Hydras are ordinarily sessile, being attached by a disc-like foot to some solid support or to the surface film, from which they often hang suspended. But at times of abundance (and under conditions that are not at present well understood) they become detached and drift about in the water. A hydra of a brick-red color swarms about the outlet of Little Clear Pond at Saranac Inn, N. Y., in early summer, and drifts down the out-flowing stream, often in such abundance that the water is tinged with red. The young trout in hatching ponds through which this stream flows, neglect their regular ration of ground liver, and feed exclusively upon the hydras, so long as the abundance continues. The hydras play fast-and-loose in the stream, attaching themselves when they meet with some solid support, and then loosening and drifting again.

Clear, sunlit pools are the favorite haunts of hydras, and the early summer appears to be the time of their maximum abundance. They attach themselves mainly to submerged stems and leaves, and to the underside of floating duckmeat. They feed upon lesser animals which abound in the plancton, and, multiplying rapidly by a simple vegetative process of budding with subsequent detachment, they become numerous when plancton abounds. Kofoid ('08) found a maximum number of 5335 hydras per cubic meter of water in Quiver Lake during a vernal plancton pulse in 1897.

Fresh-water sponges grow abundantly in the margins of lakes and pools and in clear, slow-flowing streams. They are always sessile upon some solid support. In sunlight they are green, in the shade they grow pale. The species that branch out in slender finger-like processes are most suggestive of plants in both form and color, but even the slen-

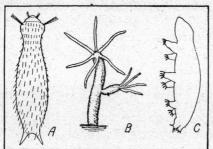

derest sponge is more massive than any plant body; and when one looks closely at the surface he sees it roughened all over with the points of innumerable spicules, and sees open osteoles at the tips. By these signs sponges of whatever form or color are easily recognized.

FIG. 74. Three simple metazoans of isolated structural types.

A, a scruff back, *Chætonctus; B, Hydra,* bearing a bud; *C,* a tardigrade, *Macrobiotus.*

The commonest sponges are low encrusting species that grow outspread over the surfaces of logs and timbers. When, in early summer, one overturns a floating log that has been long undisturbed he may find it dotted with young sponges, growing as little yellow, circular, fleshy discs, bristling with spicules, and each with a large central osteole. Later they grow irregular in outline, and thicker in mass. Toward the end of their growing season they develop statoblasts or gemmules (winter-buds) next to the substratum (see fig. 164 on p. 264), and then they die and disintegrate. So our fresh-water sponges are creatures of summer, like daffodils.

All sponges are aquatic, and most of them are marine. Only the fresh-water forms produce statoblasts, and live as annuals.

In figure 74 we show two other simple metazoans (unrelated to Hydra and of higher structural rank

FIG. 75. A semi-columnar sponge from the Fulton Chain of Lakes near Old Forge, N. Y. Half natural size. Photo. kindly loaned by Dr. E. P. Felt of the N. Y. State Museum.

than the sponges) that during the history of systematic zoology, have been much bandied about among the groups, seeking proper taxonomic associates. *Chætonotus* often appears on the side of an aquarium jar gliding slowly over the surface of the glass as a minute oblong white speck. It is an inhabitant of water containing plant infusions, and an associate of Paramecium which to the naked eye it somewhat resembles.

Macrobiotus may be met in the same way and place, but less commonly. It may also be taken in plancton; but its favorite habitat appears to be tangles of water-plants, over whose stems it crawls clumsily with the aid of its four pairs of stub-by strong-clawed feet. It also inhabits the most temporary pools, even rainspouts and stove urns, and is able to withstand dessi-cation.

Chætonotus is probably most nearly related to the Rotifers; Macro-biotus, to the mites.

Bryozoans — The Bryozoans or "moss animals" (called also Polyzoans) are colonial forms that are very common in fresh water. They grow always in sessile colonies, which have a more or less plant-like mode of branching. Their fixity in place, their spreading branches and the brownish color of the test they secrete give the commoner forms

FIG. 76. Bryozoan colonies, slightly en-larged; a dense colony of *Plumatella* on a grass-stem; a beginning colony on a leaf (above); and a loosely grown colony of *Fredericella*.

an aspect enough like minute brown creeping water mosses to have suggested the name. The individ-ual animals (zooids) of a colony are minute, requir-ing a pocket lens for their examination, but the colonies are often large and conspicuous. Two of the commoner genera are shown in figure 76, natural size. These may be found in every brook or pond, growing in flat spreading colonies on leaves or pieces of bark or stones. Often a flat board that has long been floating on the water, if overturned, will show a complete and beautiful tracery of entire colonies outspread upon the surface. New zooids are produced by budding. The buds remain permanently attached, each at the tip of a branch. With growth in length and the formation of a tough brownish cuticle over every portion except the ends, the skeleton of the colony develops. This skeleton is what we see when we lift the leaf from the water and look at the colony—brown, branching tubes, with a hole in the end of each branch. Nothing that looks like an animal is visible, for the zooids which are very sensitive and very delicate have all withdrawn into shelter. They

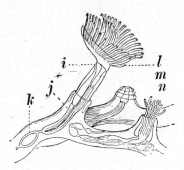

FIG. 77. Three zooids of the bryozoan, *Plumatella*, magnified.

l, expanded; *m*, retracted; *n*, partly retracted; *i*, anus; *j*, intestine; *k*, developing statoblast.

suddenly disappear on the slightest disturbance of the water, and only slowly extend again.

If we put a leaf or stone bearing a small colony into a glass of water and let it stand quietly for a time the zooids will slowly extend themselves, each unfolding a beautiful crown of tentacles. There are few more beautiful sights to be witnessed through a lens than the blossoming out of these delicate transparent, flower-like, crowns of tentacles from the tips of the apparently lifeless branches of a populous colony. They unfold from each bud, like a whorl of slender petals and slowly

extend their tips outward in graceful curves.   Then one
sees a mouth in the midst of the tentacles, and water-
currents set up by the lashing of the cilia which cover

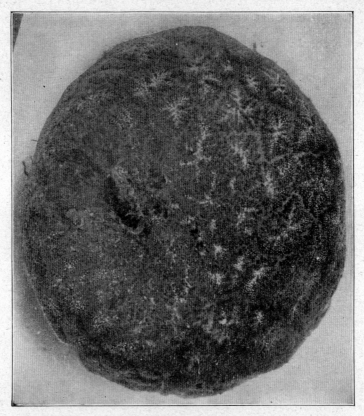

FIG. 78.   A colony of Pectinatella, one-half natural size.   Note the
distribution of buds in close groups over the surface.   The large
hole marks the location of the stick around which the colony grew.

them.   A close examination with the microscope will
reveal in each zooid the usual system of animal organs.
The alimentary canal is U-shaped its two openings
being near together at the exposed end of the body.

Several Bryzoans secrete a gelatinous covering instead of a solid tube, and the colonies become invested in a soft transparent matrix. *Pectinatella* (fig. 78) is one of these. It grows in large, more or less spherical colonies, often resembling a muskmelon in size, shape and superficial appearance. It is a not uncommon inhabitant of bayous and ditches and slow-flowing streams. It grows in most perfect spherical form when attached to a rather small twig. The clustered zooids form grayish rosettes upon the surface of the huge translucent sphere. Late in the season when statoblasts appear the surface becomes thickly besprinkled with brown. Still later, after the zooids have died, and the statoblasts have been scattered the supporting gelatin persists, blocks and segments of it, derived from disintegrating colonies, now green from an overgrowth of algæ, are scattered about the shores.

There are but a few genera of fresh-water bryozoans —some six or seven—and *Plumatella* is much the commonest one. Plumatella and allied forms grow in water pipes. They gather in enormous masses upon the sluiceways and weirs of water reservoirs. They sometimes cover every solid support with massive colonies of interlaced and heaped-up branches. Thus they form an incrusting layer thick enough to be removed from flat surfaces with shovels. Its removal is demanded because the bryozoans threaten the potability of the water supply. They do no harm while living and active, but when with unfavorable conditions they begin to die, their decomposing remains may befoul the water of an entire reservoir.

Cristatella is a flat, rather leech-shaped form that is often found on the under side of lily pads. It is remarkable for the fact that the entire colony is capable of a slow creeping locomotion. The zooids act together as one organism.

The free-living flatworms abound in most shoal fresh
waters. Some live in shallow pools; others in lakes
and rivers, others in spring-fed brooks. They gather
on the under sides of stones, sticks and trash, and con-
ceal themselves amid vegetation, usually shunning
the light. They are often collected unnoticed, and
crawl at night from cover and lie outspread upon the

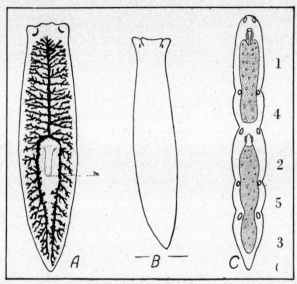

FIG. 79. Flatworms.

A, diagram of a planarian, showing food cavity; *M*, mouth at end of cylindric pharynx, directed
downward underneath the body; *B*, *Dendrocœlum*; *C*, a chain of five individuals of *Stenos-
tomum* formed by automatic division of the body, (after Keller). Note the anterior position
of the mouth and the unbranched condition of the alimentary canal in this *Rhabdocœle* type.

sides of our aquaria. We may usually find the larger
species by lifting stones from a stream bed or a lake
shore, and searching the under side of them.

Flatworms are covered with vibratile cilia and travel
from place to place with a slow gliding motion. They
range in length from less than a millimeter to several
centimeters. The smaller among them are easily mis-

taken for large ciliate protozoans, if viewed only with the unaided eye; but under the microscope the alimentary canal and other internal organs are at once apparent. They are multicellular and have little likeness to any infusoria, save in the ciliated exterior. Most members of the group are flattened, as the common name suggests, but a few are cylindric, or even filiform. A few are inclined to depart from shelter and to swim in the open water, especially at time of abundance. Kofoid ('08) found them in the channel waters of the Illinois River in average numbers above 100 per cubic meter, with a maximum record of 19250 per cubic meter.

The large flatworms resemble leeches somewhat in form of body, but they have more of a head outlined at the anterior end. They lack the segmentation of body and the attachment discs of leeches, and their mode of locomotion is so very different they are readily distinguished. They do not travel by loopings of the body as do leeches, but they glide along steadily, propelled by invisible cilia.

The most familiar flatworms are the planarians: soft and innocuous-looking little carnivores, having the mouth opening near the midventral surface of the body, and the food-cavity spreading through the body in three complexly ramifying branches. They are often brightly colored, mottled white, or brick red, or plumbeous, and they have a way of changing color with every full meal; for the branched alimentary canal fills, and the color of the food glows through the skin in the more transparent species. The eggs of planarians are often found in abundance on stones in streams in late summer. They are inclosed in little brownish capsules, of the size and appearance of mustard seeds, and each capsule is raised on a short stalk from the surface of the stone. Increase is also by automatic transverse division of the body, the division plane lying close behind the mouth.   When a new head has been shaped on the tail-piece, and a new tail on the head-piece, and two capable organisms have been formed, then they separate.   In some of the simple (Rhab-docœle) flatworms the body divides into more than two parts simultaneously and thus chains of new individuals arise (fig. 79 *c*).

Thread-worms or *Nematodes*, abound in all fresh waters, where they inhabit the ooze of the bottom, or thick masses of vegetation.   They are minute, color-less, unsegmented, smoothly-contoured cylindric worms rarely more than a few millimeters long.   The tail end is usually sharply pointed.   The mouth is terminal at the front end of the body, and is surrounded by a few short microscopic appendages.   Within the mouth cavity there are often little tooth-like appendages. The alimentary canal is straight and cylindric and unappendaged, and the food is semifluid organic sub-stances.

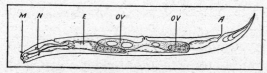

FIG. 80.   Diagram of a Nematode worm.

m, mouth;   *n*, nerve ring;   *e*, alimentary canal;   *ov*,
ov, ovaries;   *a*, anus.   (After Jagerskiold).

We can hardly collect any group of pond-dwellers without also collecting nematodes.   They may occupy any crevice.   They slip in between the wing-pads of insect nymphs, and into the sheaths of plant stems. When we disturb the trash in the bottom of our collect-ing dish, we see them swim forth, with violent swings and reversals of the pliant body.   They may easily be picked up with a pipette.

Oligochetes—Associated with the nematodes in the trash and ooze, there is a group of minute bristle-bearing worms, the naiads (Family Naidæ), similar in slenderness and transparency of body, but very different on close examination; for the body in Nais is segmented, and each segment is armed with tufts of bristles of variable length and form. There are many common members of this family. Besides the graceful Nais shown in our figure there is *Chætogaster*, which creeps on its dense bristle-clusters as on feet. There is *Stylaria* with a long tongue-like proboscis. There is *Dero* that lives at the surface in a tube of some floating plant stuffs, such as seeds (fig. 82) or Lemna leaves, slipping in and out or changing ends in the tube with wonderful celerity; and there are many others.

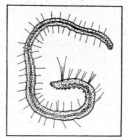

Fig. 81. Nais. (after Leunis)

Dero bears usually two pairs of short gill-lobes at the posterior end of the body.

All these naiads reproduce habitually by automatic division of the body, which when in process of development, forms chains of incompletely formed individuals, as in certain of the flatworms before described.

Another group of Oligochetes is represented by *Tubifex* and its allies. These dwell in the bottom mud, living in stationary tubes, which are in part burrows, and in part chimneys extended above the surface. The worms remain anchored in these and extend their lithe bodies forth into the water. On disturbance they vanish instantly, retreating into their tubes. They are often red in color, and when thickly associated, as on sludge in the bed of some polluted pool, they often cover the bottom as with a carpet of a pale mottled reddish color.

FIG. 82.   Dero, in its case made of floating seeds.

Aquatic earthworms, more like the well-known terrestrial species, burrow deeply into the mud of the pond bottom.

Other worms occur in the water in great variety; we have mentioned only a few of the commonest, and those most frequently seen.   There are many parasitic worms that appear in the water for only a brief period of their lives: hair-worms (*Gordius*, etc.), which are freed from the bodies of insects and other animals in which they have developed; these often appear in watering troughs and were once widely believed to have generated from horse-hairs fallen into the water. There are larval stages (*Cercaria*) of Cestodes and others, found living in the water for only a brief interval of passage from one host animal to another.   There are smaller groups also like the Nemertine worms, sparingly represented in fresh-water; for information concerning these the reader is referred to the larger textbooks of zoology.

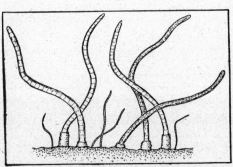

FIG. 83.   Tubifex in the bottom mud.

Leeches—The leeches constitute a small group whose members are nearly all found in fresh-water. They occur under stones and logs, in water-weeds or bottom mud, or attached to larger animals. The body is always depressed, and narrowed toward the ends, more abruptly toward the posterior end where a strong sucker is developed. The front end is more tapering and neck-like, and very pliant. There is no distinct head, but at the front is a sort of cerebral nerve ring and there are rudimentary eyes in pairs, and surrounding the mouth is a more or less well-developed anterior sucker. The great pliancy of the muscular body, the presence of the two terminal suckers, and the absence of legs or other appendages determine the leech's mode of locomotion. It ordinarily crawls about by a series of loopings like a "measuring worm," using the suckers like legs for attachment. The more elongate leeches swim readily with gentle undulations of the ribbon-like body. The shorter broader forms hold more constantly with the rear sucker to some solid support, and when detached tend to curl up ventrally like an armadillo.

Leeches range in size from little pale species half an inch long when grown, to the huge blackish members of the horse-leech group (*Hæmopis*) a foot or more in length. Many of them are beautifully colored with soft green and yellow tints. The much branched alimentary canal, when filled with food, shows through the skin of the more transparent species in a pattern that is highly decorative.

Leeches eat mainly animal food. They are parasites on large animals or foragers on small animals or scavengers on dead animals. Very commonly one finds the parasites attached to the thinner portions of the skins of turtles, frogs, fishes and craw-fishes. There is no group in which the boundary between predatory and parasitic habits is less distinct than in this one; many leeches will make a feast of vertebrate blood, if occasion offers, or in absence of this will swallow a few worms instead.

FIG. 84. A clepsine leech (*Placobdella rugosa*), overturned and showing the brood of young protected beneath the body. (From the senior author's *General Biology*).

The mouth of leeches is adapted for sucking, in some cases it is armed for making punctures, as well: hence the food is either more or less fluid substances like blood or the decomposing bodies of dead animals, or else it consists of the soft bodies of animals small enough to be swallowed whole.

The eggs of leeches are cared for in various ways: commonly one finds certain of them in minute packets, attached to stones. Others (*Hæmopis*, etc.) are stored in larger capsules and hidden amid submerged trash. Others are sheltered beneath the body of the parent, and the young are brooded there for a time after hatching, as shown in the accompanying figure. Nachtrieb (12) states that they are so carried "until the young are able to move about actively and find a host for a meal of blood."

Leeches are doubtless fed upon by many carnivorous animals. They are commonly reported to be taken freely by the trout in Adirondack waters. In Bald Mountain Pond they swim abundantly in the open water.

The Rotifers constitute a large group of minute animals, most characteristic of fresh-water. They abound in all sorts of situations, and present an extraordinary variety of forms and habits. Their habits vary from ranging the open lake to dwelling symbiotically within the tissues of water plants; from sojourning in the cool waters of perennial springs, to running a swift course during the temporary existence of the most transient pools. They even maintain themselves in rainspouts and stone urns, where they become desiccated with evaporation between times of rain.

Rotifers are mainly microscopic, but a few of the larger forms are recognizable with the unaided eye. Often they become so abundant in pools as to give to the water a tinge of their own color. Grouped together in colonies they become rather conspicuous. The spherical colonies of *Cono-*

Fig. 85. Three colonies of the rotifer, *Conochilus*, attached to the tips of leaves of the pond-weed, *Nais*.

chilus when attached to leaf-tips, as in the accompanying picture, present a bright and flower-like appearance. Entire colonies often become detached, and then they go bowling along through the water, in a most interesting fashion, the individuals jostling each other as they stand on a common footing, and all merrily waving their crowns of cilia in unison. Often a little roadside pool will be found teeming with the little white rolling spheres, that are quite large enough to be visible to the unaided eye.

Melicerta is a large sessile rotifer that lives attached to the stems of water-plants and when undisturbed protrudes its head from the open end of the tube, and unfolds an enormous four-lobed crown of waving cilia. It is a beautiful creature. Our picture shows the cases of a number of Melicertas, aggregated together in a cluster, one case serving as a support for the others.

The crown of cilia about the anterior end of the body is the most characteristic structure possessed by rotifers. It is often circular, and the waving cilia give it an aspect of rotation, whence the group name. It is developed in an extraordinary variety of ways as one may see by consulting in any book on rotifers the figures of such as *Stephanoceros*, *Floscularia*, *Synchæta*, *Trochosphæra* and *Brachionus*.

The cilia are used for driving food toward the mouth that lies in their midst, and for swimming. Most of the forms are free-swimming, and many alternately creep and swim.

FIG. 86. Two clusters of rotifers (*Melicerta*), the upper but little magnified. Only the cases (none of the animals) appear in the photographs.

Brachionus (fig. 87) shows well the parts commonly found in rotifers. The body is inclosed in a lorica or shell that is toothed in front and angled behind. From its rear protrudes a long wrinkled muscular "foot," with two short "toes" at its tip. This serves for creeping. The lobed crown of cilia occupies the front. Behind the quad-

rangular black eyespot in the center of the body
appears the food communicating apparatus (mastax),
below which lie ovaries and alimentary canal.   Any
or all the external parts may be wanting in certain

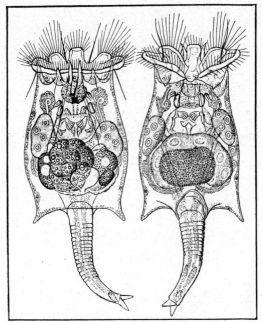

FIG. 87.   A rotifer (*Brachionus entzii*) in dorsal and ventral views.   (After Francé).

rotifers.   The smaller and simpler forms superficially
resemble ciliate infusoria, but the complex organization
shown by the microscope will at once distinguish them.
    Rotifers eat micro-organisms smaller than them-
selves.   They reproduce by means of eggs, often
parthenogenetically.   The males in all species are
smaller than the females and for some species males
are not known.

Molluscs—A large part of the population of lake and river beds, shores, and pools is made up of molluscs. They cling, they climb, they burrow, they float—they do everything but swim in the water. They are predominantly herbivorous, and constitute a large proportion of the producing class among aquatic animals. Two great groups of molluscs are common in fresh water, the familiar groups of mussels and snails.

Fresh-water mussels (clams, or bivalves) abound in suitable places, where they push through the mud or sand with their muscular protrusible foot, and drag the shell along in a vertical position leaving a channel-

Fig. 88. A living mussel, *Anodonta*, with foot retracted and shell tightly closed. A copious growth of algæ covers the portion of the shell that is exposed above the mud in locomotion: the remainder is buried in oblique position with the foot projecting still more deeply into the mud.

like trail across the bottom.   They feed on micro-organisms.

The two commonest sorts of fresh-water mussels are roughly distinguished by size and reproductive habits thus: *Unios* and their allies are large forms that have pearly shells and that live mainly in large streams and lake borders. They produce enormous numbers of young, and use mostly the outer gill for a brood chamber. They cast the young forth while still minute as *glochidia*, to become attached to and temporarily parasitic on fishes. The relations of these larval glochidia with the fishes will be discussed in chapter V.

The lesser mussels (family Sphæridæ) dwell in small streams and pools and in the deeper waters of lakes. Their shells are not pearly. They produce but a few young at a time and carry these until of large size, using the inner gill for a brood-pouch. The stouter species, half an inch long when grown, burrow in stream-beds like the unios. The slenderer species climb up the stems of plants by means of their excessively mobile adhesive and flexible foot. On this foot the dainty white mussel glides like a snail or a flatworm, up or down, wherever it chooses.

Snails are as a rule more in evidence than are mussels, for they come out more in the open. They clamber on plants and over every sort of solid support. They hang suspended from the surface film, or descend therefrom on strings of secreted mucus. They traverse the bottom ooze. We overturn a floating board and find dozens of them clinging to it, and often we find a filmy green mass of floating algæ thickly dotted with their black shells.

They eat mainly the soft tissues of plants, and microorganisms in the ooze covering plant stems. A ribbonlike rasp (*radula*) within the mouth drawn back and forth across the plant tissue scrapes it and comminutes it for swallowing. Because snails wander constantly and feed superficially without, as a rule, greatly altering the form and appearance of the larger plants on which they feed, their work is little noticed; yet they consume vast quantities of green tissue and dead stems. The commoner pond snails lay their eggs in oblong gelatinous clumps that are outspread upon the surfaces of leaves and other solid supports. Other snails are viviparous.

The two principal groups of fresh-water snails may roughly be distinguished as (1) operculate snails which live mainly upon the bottom in larger bodies of water, and have an operculum closing the aperture of their shell when they retreat inside, and which breathe by

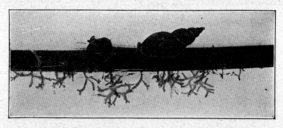

Fig. 89. Two pond snails (*Limnæa palustris*) foraging on a dead stem that is covered with a fine growth of the alga, *Chætophora incrassata*.

means of gills: (2) pulmonate snails, which most abound in vegetation-filled shoals, breathe by means of a simple lung (and come to the surface betimes, to refill it with air) and have no operculum.

The snails we oftenest see are members of three genera of the latter group: *Limnæa*, shown in the accompanying figure, having a shell with a right-hand spiral and a slender point; *Physa*, having a shorter spiral, twisted in the opposite way, and *Planorbis*, shown in fig. 65 on p. 155, having a shell coiled in a flat spiral. *Ancylus* is a related minute limpet-shaped snail, having a widely open shell that is not coiled in a spiral. Its flaring edges attach it closely to the smooth surfaces of plant stems or of stones.

ARTHROPODS

We come now to that great assemblage of animals which bear a chitinous armor on the outside of the body, and, as the name implies, are possessed of jointed feet. This group is numerically dominant in the world today on sea and land. It is roughly divisible into three main parts; crustaceans, spiders and insects. The crustaceans are the most primitive and the most wide-spread in the water-world; so with them we will begin.

The Crustaceans include a host of minute forms, such as the water fleas and their allies, collectively known as *Entomostraca*, and a number of groups of larger forms, such as scuds, shrimps, prawns and crabs, collectively known as the higher Crustacea or *Malacostraca*. A few of the latter (crabs, sow-bugs, etc.) live in part on land, but all the groups are predominately aquatic, and the Entomostraca are almost wholly so.

The Entomostraca are among the most important animals in all fresh waters. They are perhaps the chief means of turning the minute plant life of the waters into food for the higher animals. They are themselves the chief food of nearly all young fishes.

There are three groups of Entomostraca, so common and so important in fresh water, that even in this brief discussion we must distinguish them. They are: *Branchiopods, Ostracods* and *Copepods.*

The Branchiopods, or gill-footed crustaceans, have some portion of the thoracic feet expanded and lamelliform, and adapted to respiratory use. The feet are moved with a rapid shuttle-like vibration which draws the water along and renews the supply of oxygen. The largest of the entomostraca are members of this group; they are very diverse in form.

The fairy shrimp, shown in the accompanying figure, is one of the largest and showiest of Entomostraca. It is an inch and a half long and has all of the tints of the rainbow in its transparent body. It appears in spring in rainwater pools and is notable for its rapid growth and sudden disappearance. It runs its rapid course while the pools are filled with water, and lays its eggs and dies before the time of their drying up. The eggs settle to the bottom and remain dormant, awaiting the return of favorable season. The animal swims gracefully on its back with two long rows of broad, thin, fringed, undulating legs uppermost, and its forked tail streaming out behind, and its rich colors fairly shimmering in the light.

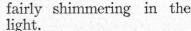

FIG. 90. The Fairy Shrimp, *Chirocephalus* (after Baird).

Of very different appearance is the related mussel-shrimp (*Estheria*), which has its body and its long series of appendages inclosed in a bivalve shell. Swimming through the water, it looks like a minute clam a centimeter long, traveling in some unaccountable fashion; for its legs are all hidden inside, and nothing but the translucent brownish shell is visible. This shell is singularly clam-like in its concentric lines of growth on the surface and its umbones at the top. This, in America, is mainly Western and Southern in its distribution, as is also *Apus*, which has a single dorsal shell or carapace, widely open below and shaped like a horseshoe crab.

These large and aberrant Branchiopods are all very local in distribution and of sporadic occurrence. As the seasons fluctuate, so do they. But they are so unique in form and appearance that when they occur they will hardly escape the notice of the careful observer of water life.

Water-fleas—The most common of the Branchiopods are the water-fleas (order Cladocera) such as are shown in outline in figure 91. These are smaller, more transparent forms, having the body, but not the head, inclosed in a bivalve shell. The shell is thin, and finely reticulate or striated or sculptured, and often armed with conspicuous spines. The post-abdomen is thin and flat, armed with stout claws at its tip and fringed with teeth on its rear margin, and it is moved in and out between the valves of the shell like a knife blade in its handle. The pulsating heart, the circulating blood, the contracting muscles, and the vibrating gill-feet all show through the shell most clearly under a microscope. Hence these forms are very interesting for laboratory study, requiring no preparation other than mounting on a slide.

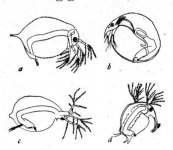

FIG. 91. Water-fleas
a, Daphne; *b*, Chydorus; *c*, Simocephalus; *d*, Bosmina. Note the "proboscis."

Some water-fleas, like *Simocephalus*, shown in figures 91 and 92 swim freely on their backs, in which position gravity may aid them in getting food into their mouths. When the swimming antennæ are developed to great size, as in *Daphne* (fig. 91*a*), the strokes are slow and progress is made through the water in a series of jumps. When the antennæ are shorter, as in *Chydorus* (fig. 91*b*), their strokes are more rapidly repeated, and progression steadier.

The Cladocerans are abundant plancton organisms throughout the summer season. They forage at a little depth by day, and rise nearer to the surface by night.

The food of water-fleas is mainly the lesser green algæ and diatoms. They are among the most important

herbivores of the open water. They are themselves
important food for fishes.

The importance of water fleas in the economy of
water is largely due to their very rapid rate of reproduction. During the summer season broods of eggs suc-

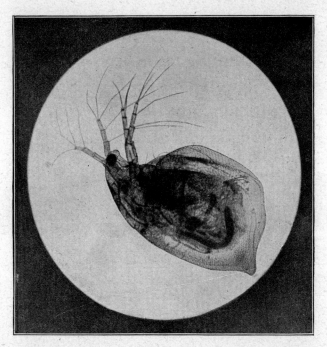

FIG. 92. A water-flea (*Simocephalus vetulus*) in its ordinary
swimming position. Note the striated shell, and the alimentary canal, blackish where packed with food-residue in
the abdomen.

cessively appear in the chamber enclosed by the shell
on the back of the animal (see figure 93) at intervals
of only a few days. The young develop rapidly and
are themselves soon producing eggs. In *Daphne pulex*,
for example, it has been calculated that the possible

progeny of a single female might reach the astounding number of 13,000,000,000 in sixty days.

The Ostracods are minute crustaceans, averaging perhaps a millimeter in length, having the head, body and appendages all inclosed in a bivalve shell. The shell is heavier and less transparent than that of the water fleas. It is often sculptured, or marked in broad patterns

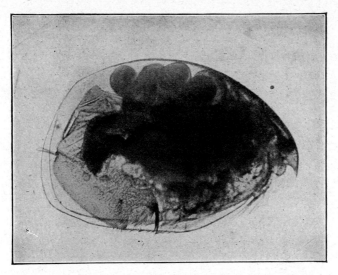

Fig. 93. One of our largest water-fleas, *Eurycerus lamellatus*, twenty times natural size. Note the eggs in the brood chamber on the back. Note also the short beak and the broad post-abdomen (shaped somewhat like a butcher's cleaver) by which this water-flea is readily recognized.

with darker and lighter colors. The inclosed appendages are few and short, hardly more than their tips showing when in active locomotion. There are never more than two pairs of thoracic legs. The identification of ostracods is difficult, since, excepting in the case of strongly marked forms, a dissection of the animal from its shell is first required.

Some Ostracods are free-
swimming (species of *Cypris*,
etc.) and some (*Notodromas*)
haunt the surface in sum-
mer; but most are creeping
forms that live among
water plants or that burrow

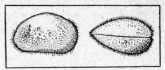

FIG. 94. An Ostracod (*Cypris virens*), lateral and dorsal views, (after Sharpe.)

in the bottom ooze.   In pools where such food as algæ
and decaying plants abound Ostracods frequently
swarm, and appear as a multitude of moving specks
when we look down into the still water.

Relict pools in a dry summer are likely to be found
full of them.   Both sexes are constantly present in
most species of Ostracods, but a few species are represented by females only, and reproduce by means of
unfertilized eggs.

The Copepods are the perennial entomostraca of open
water.   Summer and winter they are present.   Three
of the commonest genera are shown in figure 95, together with a nauplius—the larval form in which the
members of this group hatch from the egg.   Nothing is
more familiar in laboratory aquaria than the little
white *Cyclops* (fig. 96, swim-
ming with a jerky motion,
the female carrying two
large sacs of eggs.

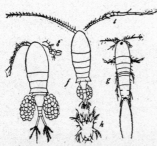

FIG. 95.   Common copepods

e, Cyclops; *f*, Diaptomus; *g*, Canthocamptus; *h*, a nauplius (larva) of Cyclops.
Figures *e* and *f* show females bearing egg
sacs, while the detached antenna at the
right shows the form of that appendage
in the male.

A more or less pear-shaped
body tapering to a bifurcate
tail at the rear, a single
median eye and a pair of
large swimming antennæ at
the front, and four pairs of
thoracic swimming feet
beneath, characterize the
members of this group.

The species of *Diaptomus* are remarkable for having usually very long antennæ and often a very lively red color. Sometimes they tinge the water with red, when present in large numbers.

Copepods feed upon animals plancton and algæ, especially diatoms. They are themselves important food for fishes, especially for young fishes.

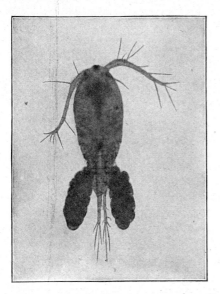

FIG. 96. A female Cyclops, with eggs.

The higher crustacea, (Malacostraca) are represented in our fresh waters by four distinct groups, all of which agree in having the body composed of twenty segments that are variously fused together on the dorsal side, each, except the last, bearing (at least during development) a pair of appendages. Of these segments five belong to the head, eight to the thorax and the remainder to the abdomen. *Mysis* (fig. 97) is the sole representative of the most primitive of these groups, the order Mysidacea. Its thoracic appendages are all biramous and undifferentiated; and still serve their primal swimming function. Mysis lives in the open waters of our larger lakes, in their cooler depths. It is a delicate transparent creature half an inch long.

The Scuds (order Amphipoda) are flattened laterally, and the body is arched. The thoracic legs are adapted for climbing, and the abdominal appendages for swimming and for jumping. The body is smooth and pale; often greenish in color. The scuds are quick and active. They dart about amid green water-weeds, usually keeping well to shelter, and they swim freely and rapidly when disturbed. In figure 98 are shown three species that are common in the eastern United States.

The scuds are herbivores, and they abound among green water plants everywhere. They are of much importance as food for fishes. They are hardy, and capable of maintaining themselves under stress of

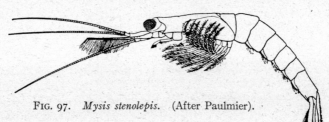

FIG. 97. *Mysis stenolepis.* (After Paulmier).

competition. They carry their young in a pectoral broodpouch until well developed; and altho they are not so prolific as are many other aquatic herbivores, yet they have possibilities of very considerable increase, as is shown by the following figures for *Gammarus fasciatus*, taken from Embody's studies of 1912:

Reproductive season at Ithaca, Apr. 18th to Nov. 3d, includes 199 days.

Average number of eggs laid at a time 22. Egg laying repeated on an average of 11 days.

Age of the youngest egg-laying female 39 days: number of her eggs, 6.

Possible progeny of a single pair 24221 annually.

Asellus is the commonest representative of the order Isopoda; broad, dorsally-flattened crustaceans of some-

what larger size, that live sprawling in the mud of the
bottom in trashy pools. Their long legs and hairy
bodies are thickly covered with silt. Two pairs of
thoracic legs are adapted for grasping and five pairs for
walking, and the appendages of the middle abdominal
segments are modified to serve for respiration. Asellus
feeds on water-cress and on other soft plants, living and
dead, are found in the bottom ooze. It reproduces
rapidly, and, in spite of cannibal habits when young,

FIG. 98. Three common Amphipods.
A, Gammarus limnæus; B, Gammarus fasciatus; C, Eucrangonyx gracilis.
(Photo by G. E. Embody).

often becomes exceedingly abundant. An adult female
of *Asellus communis* produces about sixty eggs at a
time and carries them in a broodpouch underneath her
broad thorax during their incubation. There is a new
brood about every five or six weeks during the early
summer season.

Both this order and the preceding have blind
representatives that live in unlighted cave waters, and
pale half-colored species that live in wells.

The crawfishes are the commonest inland representatives of the order Decapoda. These have the thoracic

segments consolidated on the dorsal side to form a hard carapace, and have but five pairs of walking legs (as the group name indicates), the foremost of these bearing large nipper-feet. This group contains the largest crustacea, including all the edible forms, such as crabs, lobsters, shrimps, and prawns, most of which are marine. Southward in the United States there are fresh-water prawns (*Palæmonetes*) of some importance as fish food.

The eggs of crawfishes are carried during incubation, attached to the swimmerets of the abdomen, and the young are of the form of the adult when hatched. They cling for a time after hatching to the hairs of the swimmerets by means of their little nipper-feet, and are carried about by the mother crawfish.

Crawfishes are mainly carnivorous, their food being smaller animals, dead or alive, and decomposing flesh. In captivity they are readily fed on scraps of meat. Southward, an omnivorous species is a great depredator in newly planted fields of corn and cotton. Hankinson ('08) reports that the crawfishes "form a very important if not the chief food of black bass, rock bass, and perch" in Walnut Lake, Michigan.

FIG. 99. *Asellus aquaticus*, (x2 after Sars)

Spiders and *Mites* are nearly all terrestrial. Of the true spiders there are but a few that frequent the water. Such an one is shown in the initial cut on page 158. This spider is conspicuous enough, running on the surface of the water, or descending beneath, enveloped in a film of air that shines like silver; but neither this nor any other true spider is of so great importance in the economy of the water as are many other animals that are far less conspicuous. In habits these do not differ materially from their terrestrial relatives.

Of mites there is one rather small family (Hydrach-nidæ) of aquatic habits. These water-mites are minute, mostly rotund (sometimes bizarre) forms with unsegmented bodies, and four pairs of long, slender, radiating legs. One large species (about the size of a small pea) is so abundant in pools and is so brilliant red in color that it is encountered by every collector. Others, tho

FIG. 100. An overturned female crawfish (*Cambarus bartoni*), showing the eggs attached to the swimmerets (four thoracic legs broken off).

smaller, are likewise brilliant with hues of orange, green, yellow, brown and blue, often in striking patterns.

Water-mites, even when too small to be distinguished easily by their form from ostracods or other minute crustacea are easily distinguished by their manner of locomotion. They swim steadily, in one position; not in the jerky manner of the entomostraca. The strokes of their eight hair-fringed swimming feet come in such rapid succession that the body is moved smoothly forward. A few water-mites that dwell in the open water of lakes are transparent, like other members of open-water plancton.

Water-mites are nearly all parasitic: they puncture the skin and suck the blood of larger aquatic animals. Certain of them are common on the gills of mussels: others on the intersegmental membranes of insects.

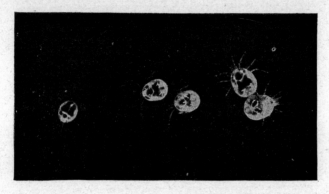

FIG. 101.   Water mites of the genus *Limnochares*

Nothing is more common than to find clusters of red mites hanging conspicuously at the sutures of back-swimmers and other water insects.

Many mites lay their minute eggs on the surface of the leaves of water plants. Their young on hatching have but three pairs of legs.

INSECTS

This is the group of animals that is numerically dominant on the earth today. There are more known species of insects than of all other animal groups put together. The species that gather at the water-side give evidence, too, of most extraordinary abundance of individuals. Who can estimate the number of midges in the swarms that hover like clouds over a marsh, or the number of mayflies represented by a windrow of cast skins fringing the shore line of a great lake? The world is full of them. Like other land animals they are especially abundant about the shore line, where conditions of water, warmth, air and light, favor organic productiveness.

Nine orders of insects (as orders are now generally recognized) are found commonly in the water. These are the Plecoptera or stoneflies; the Ephemerida or mayflies; the Odonata or dragonflies and damselflies; the Hemiptera or water bugs; the Neuroptera or net-winged insects; the Trichoptera or caddis-flies; the Lepidoptera or moths; the Coleoptera or beetles; and the Diptera or true flies. These, together with the Thysanura or springtails, which hop about upon the surface of the water in pools, and the Hymenoptera, of which a few members are minute egg-parasites and which, when adult, swim with their wings, represent the entire range of hexapod structure and metamorphosis. Yet the six-footed insects as a class are predominantly terrestrial. It is only a few of the smaller orders, such as the stoneflies and the mayflies, that are wholly aquatic. Of the very large orders of moths, beetles and true flies only a few are aquatic.

Aquatic insects are mainly so in their immature stages; the adults are terrestrial or aerial. Only a few adult bugs and beetles are commonly found in the

water. Other insects are there as nymphs or larvæ;
and, owing to the great change of form that is undergone

FIG. 102. The green darner dragonfly, *Anax junius;* adult and nymph
skin from which it has just recently emerged. Save for the displaced
wing cases the skin preserves well the form of the immature stage.
Photo by H. H. Knight

at their final transformation, they are very unlike the
adults in appearance. How very unlike the brilliant

adult dragonfly, that dashes about in the air on shimmering wings, is the sluggish silt-covered nymph, that sprawls in the mud on the pond bottom! How unlike the fluttering fragile caddis-fly is the caddis-worm in its lumbering case!

As with terrestrial insects, so with those that are aquatic, there are many degrees of difference between young and adult, and there are two main types of metamorphosis, long familiarly known as complete and incomplete. With complete metamorphosis a quiescent pupal stage is entered upon at the close of the active larval life, and the form of the body is greatly altered during transformation. Adults and young are very unlike. Caddis-worms, for example, the larvæ of caddis-flies, are so unlike caddis-flies in every external feature, that no one who has not studied them would think of their identity.

The caddis-fly shown in the accompanying figure is one that is very common about marshes, where its larva dwells in temporary ponds and pools. Often in early summer, the bottom will be found thickly strewn with larvæ in their lumbering cases. Then they suddenly disappear.

FIG. 103. Caddis-fly.
(*Limnophilus sp.*)

They drag their cases into the shelter of sedge clumps bordering the pools, and transform to pupæ inside them. A fortnight later they transform to adult caddis-flies, and appear as shown in figure 103, pretty soft brown insects marked with straw-yellow in a neat pattern. The larva is of the form shown in figure 104, a stocky worm-like

FIG. 104.   Caddis-worms: larvæ of *Halesus guttifer*.

creature, half soft and pale where constantly protected by the walls of the case in which it lives, and half dark colored and strongly chitinized where exposed at the ends. There are stout claws at the rear for clutching the wall of the case; there are soft pale filamentous gills arranged along the side of the abdomen, and there are three spacing tubercles upon the first segment of the abdomen for insuring that a fresh supply of water shall be admitted to the case to flow over the gills. The legs are directed forward, for

FIG. 105.   The larval case of Limnophilus, attached endwise to a submerged flag leaf, in position of transformation.

Fig. 106. End view of pupal case of Limnophilus showing silken barrier; enlarged.

readier egress from the case; they reach forth from the front end, clutching any solid support.

The larva of Limnophilus lives in the case shown in figure 105. This is a dwelling composed of flat plant fragments placed edgewise and attached to the outside of a thin silken tube.

The larva, living in this tube, clambers about over the vegetation, jerkily dragging its cumbrous case along, foraging here and there where softened plant tissues offer, and when disturbed, quickly retreating inside. It frequently makes additions to the front of its case, and casts off fragments from the rear; so it increases the diameter to accommodate its own growth.

When fully grown and ready for transformation the larva partially closes the ends, spins across them net-like barriers of silk to keep out intruders while admitting a fresh water supply.

Then it molts its last larval skin and transforms into a pupa, of the form shown in the accompanying figure, having large compound eyes, long antennæ, broad external wing-cases and copious external gills.

Fig. 107. Pupa of Limnophilus.

Then ensues a quiescent period of a fortnight or more during which great changes of form, both external and internal, take place. The stuffs that the larva accumulated and built into its body during its days of foraging, and that now lie inert in the soft white body of the pupa are being rapidly made over into the form in which they will shortly appear in the body of the dainty aerial caddis-fly. However, the pupa is not wholly inactive. By gentle undulations of its body it keeps the water flowing about its gills; and when, at the approach of final transformation, its new muscles have grown strong enough, it is seized with a sudden fit of activity. It breaks through the barred door of the case, pushes out, swims away, and then walks on the surface of the water, seeking some emergent plant stem, up which to climb to a suitable place for its final transformation. There the caddis-fly emerges, at first limp and pale, but soon becoming daintily tinted with yellow and brown, full-fledged and capable of meeting the exigencies of life in a new and wholly different environment.

FIG. 108. Pupal skins of Limnophilus, left at final molting attached to a reed above the surface of the water.

It is a marvelous change of form and habits that insects undergo in metamorphosis—especially in complete metamorphosis. Such transformations as occur in other groups are hardly comparable with it. The change from a tadpole to a frog, or from a nauplius to an adult copepod, is slight by comparison; for there is no cessation of activity, and no considerable part of the body is even temporarily put out of use. But in all the higher insects an extraordinary reversal of development occurs at the close of active larval life. The larval tissues and organs disintegrate, and return to a sort of embryonic condition, to be rebuilt in new form in the adult insect.

With incomplete metamorphosis development is more direct, there is no pupal stage, and the form of the body is less altered during transformation. Metamorphosis is incomplete in the stoneflies, the mayflies, the dragonflies and damselflies and in the water bugs. The immature stage we shall speak of as a nymph. All nymphs agree in having the wings developed externally upon the sides of the thorax. Metamorphosis is complete in all the other orders above mentioned. Their immature stage we shall call

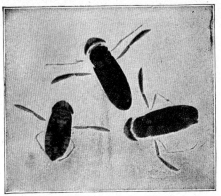

Fig. 109. Water boatmen (*Corixa*), two adults and a nymph of the same species.

a larva. All larvæ agree in having the wings developed internally: they are invisible from the outside until the pupal stage is assumed. It should be noted in passing that "complete" and "incomplete" as applied to metamorphosis are purely relative terms. There is in the insect series a progressive divergence in form between immature and adult stages, and the pupal stage comes in to bridge the widening gap between.

There is less change of form in the water bugs than in any other group of aquatic insects. The nymph of the water boatman (fig. 109) differs chiefly from the adult in the undeveloped condition of its wings and reproductive organs.

The groups of aquatic insects that are most completely given over to aquatic habits are the more generalized orders that were long included in the single Linnæan order Neuroptera (stoneflies, mayflies, dragonflies, caddis-flies, etc.)* Our knowledge of the immature stages of aquatic insects was begun by the early microscopists to whom reference has already been made in these pages: Swammerdam, Rœsel, Reaumur, and their contemporaries.† They delighted to observe and describe the developmental stages of aquatic insects, and did so with rare fidelity. After the days of these pioneers, for a long time little attention was paid to the immature stages, and descriptions of these and accounts of their habits are still widely scattered‡.

It is during their immature stages that most insects, both aquatic and terrestrial ones, are of economic importance. It is then they mainly feed and grow. It is then they are mainly fed upon. The adults of many groups eat nothing at all: their chief concern is with mating and egg-laying. Hence the study of the immature stages is worthy of the increased attention it is receiving in our own time. It will be a very long time before the life histories and habits of all our aquatic insects are made known, and there is abundant opportunity for even the amateur and isolated student of nature to make additions to our knowledge by work in this field.

---

*Under this name (we still call them *Neuropteroids*) the American forms were first described and catalogued by Dr. H. A. Hagen in his classic "*Synopsis of the Neuroptera of North America.*" (Washington, 1861). Bugs, beetles, moths and flies have received corresponding treatment in systematic synopses of their respective orders, only the adult forms being considered.

†Much of the best of the work of these pioneers has been gathered from their ancient ponderous and rather inaccessible tomes, and translated by Professor L. C. Miall, and reprinted in convenient form in his "Natural History of Aquatic Insects" (London, 1895).

‡The completest available accounts of the life histories and habits of North American aquatic insects have been published by the senior author and his collaborators in the Bulletins 47, 68, 86 and 124 of the New York State Museum.

The stoneflies (order Plecoptera) are all aquatic. They live in rapid streams, and on the wave-washed rocky shores of lakes. They are among the most generalized of winged insects. The adults are flat-bodied inconspicuous creatures of secretive habits.

Little is seen of them by day, and less by night, except when some brilliant light by the waterside attracts them to flutter around it. The colors are obscure, being predominantly black, brown or gray; but the diurnally-active foliage inhabiting chloroperlas are pale green. They take wing awkwardly and fly rather slowly, and may often be caught in the unaided hand. They are readily picked up with

Fig. 110. An adult stonefly, *Perla immarginata.*

the fingers when at rest. The wings (sometimes aborted) are folded flat upon the back. They are rather irregularly traversed with heavy veins. The tarsi are three-jointed. This, together with the flattened head, bare skin, and long forwardly-directed antennæ, will be sufficient for recognition of members of this group.

Stonefly nymphs are elongate and flattened, and very similar to the adults in form of body. They possess always a pair of tails at the end of the body. Most of them have filamentous gills underneath the body, tho a few that live in well aërated waters are lacking these. The colors of the nymphs are often livelier than those of the adults, they being adorned with bright greens and yellows in ornate patterns.

The nymphs are mainly carnivorous. They feed upon mayfly nymphs and midge larvæ and many other small animals occurring in their haunts.

One finds these nymphs by lifting stones from water where it runs swiftly, and quickly inverting them. The nymphs cling closely to the under side of the stones, lying flat with legs outspread, and holding on by means of stout paired

FIG. 111. The nymph of a stonefly, *Perla immarginata.*
(*Photo by Lucy Wright Smith.*)

claws that are like grappling hooks. Their legs are flattened and laid down against the stone in such a way that they offer little resistance to the passing current. Stonefly nymphs are always found associated with flat-bodied Mayfly nymphs of similar form, and with greenish net-spinning caddis-worms.

The mayflies (order *Ephemerida*) are all aquatic. They live in all fresh waters, being adapted to the

greatest diversity of situations. The adults are fragile insects, having long fore legs that are habitually stretched far forward, and two or three long tails that are extended from the tip of the body backward. The wings are corrugated and fan like, but not folded, and are held vertically in repose. The hind wings are small and inconspicuous. The antennæ are minute and setaceous. The head is contracted below and the mouth parts are rudimentary. Thus, many characters serve to distinguish the mayflies from other insects and make their group one of the easiest to recognize.

Mayflies are peculiar also, in their metamorphosis. They undergo a moult after the assumption of the adult form. They transform usually at the surface of the water, and, leaving the cast-off nymphal skin floating, fly away to the trees. Body and wings are then clothed in a thin pellicle of dull grayish and usually pilose skin, which is retained during a short period of quiescence. During this period (which lasts but a few minutes in Cænis and its allies, but which in the larger forms lasts

FIG. 112. An adult mayfly, *Siphlonurus alternatus*.

one or two days) they are known as subimagos or

duns.   Then this outer skin is shed, and they come
forth with smooth and shining surfaces and brighter
colors, as imagos, fully adult, and ready for their
mating flight.   Lacking mouth parts and feeding not
at all, they then live but a few hours.

There are few phenomena
of the insect world more striking than the mating flight of
mayflies.   The adult males fly
in companies, each species
maneuvering according to its
habit, and the females come
out to meet them in the air.
Certain large species that are
concerted in their season of
appearance gather in vast
swarms about the shores of all
our larger bodies of fresh water
at their appointed time.   By
day we see them sitting
motionless on every solid support, often bending the streamside willows with their weight;
and when twilight falls we see
all that have passed their final
molt swarming in untold numbers over the surface of the
water along shore.

FIG. 113.   The nymph of the
mayfly, *Siphlonurus alternatus.*
(*Photo by Anna Haven Morgan.*)

The nymphs of mayflies are all recognizable by the
gills upon the back of the abdomen.   These are
arranged in pairs at the sides of some or all of the first
seven segments.   The body terminates occasionally in
two but usually in three long tails.   The mouth parts
are furnished with many specialties for raking diatoms
and for rasping decayed stems.   Mayfly nymphs are
among the most important herbivores in all fresh waters.

The dragonflies and damselflies (order *Odonata*) are all aquatic. The adults are carnivorous insects that go hawking about over the surfaces of ponds and meadows, capturing and eating a great variety of lesser insects. The larger dragonflies eat the smaller ones.

FIG 114. An adult damselfly, *Ischnura verticalis*, perching on the stem of a low galingale, *Cyperus diandrus*.

The form of body in the dragonflies is peculiar and distinctive. The head, which is nearly overspread by the huge eyes, is loosely poised on the apex of a narrow prothorax. The remainder of the thorax is enlarged and the wings are shifted backward upon it, and the legs forward, adapting them for perching on vertical stems.

The abdomen is long and slender. On the ventral side of its second and third segments, far removed from the openings of the sperm ducts, there is developed in the male a remarkable copulatory apparatus, that has no counterpart in any other insects. The venation of the wings, also, is peculiar, nothing like it being found in any other order.

The dragonflies hold their wings horizontally in repose. The damselflies are slender forms that hold their wings vertically (or, in *Lestes*, obliquely outward) in repose. Fore and hind wings are similar in form in the damselflies; dissimilar, in the dragonflies.

Fig. 115. A nymph of the damsel-fly, *Ischnura verticalis*.

The nymphs of the entire order are recognizable by the possession of an enormous grasping labium, hinged beneath the head. This is armed with raptorial hooks and spines, and may be extended forward to a distance several times the length of the head. It is thrust out and withdrawn with a speed that the eye cannot follow. It is a very formidable weapon for the capturing of living prey. It is altogether unique among the many modifications of insect mouth parts.

Damselfly nymphs are distinguished by the possession of three flat lanceolate gill-plates that are carried like tails at the end of the abdomen. The edges of these plates are set vertically, and they are swung from side to side with a sculling motion to aid the nymphs in swimming.

Dragonfly nymphs have their gills developed upon the inner walls of a rectal respiratory chamber, and not visible externally. Hence, the abdomen is much wider than in the damselflies. Water drawn slowly into the gill chamber through an anal orifice, that is guarded by elaborate strainers, may be suddenly expelled by the strong contraction of the abdominal muscles. Thus this breathing apparatus, also, is used to aid in locomotion. The body is driven forward by the expulsion of the water backward.

Damselfly nymphs live for the most part clambering about among submerged plants in still waters; a few

Fig. 116. The burrowing nymph of a Gomphine dragonfly, with an elongate terminal segment for reaching up through the bottom mud to the water.

cling to plants in the edges of the current, and a very few cling to rocks in flowing water. Dragonfly nymphs are more diversified in their habits. Many of them also clamber among plants, but more of them sprawl in the mud of the bottom, where they lie in ambush to await their prey. One considerable group (the Gomphines) is finely adapted for burrowing in the silt and sand of the bottom.

All are very voracious, eating living prey in great variety. All appear to prefer the largest game they are able to overpower. Many species are arrant cannibals, eating their own kind even when not starved to it. As a group they are among the most important carnivores in shoal fresh waters.

The true bugs (order *Hemiptera*) are mainly terrestrial, and have undergone on land their greatest differentiation. The aquatic ones are usually found in still waters and in the shelter of submerged vegetation. Tho comparatively few in species, they are important members of the predatory population of ponds and

FIG. 117. A giant water bug (*Benacus griseus*) clinging to a vertical surface under water, natural size.

pools. They are often present in great numbers, if not in great variety. The giant water bugs (fig. 117) are among the largest of aquatic insects. These are widely known from their habit of flying to arc lights, falling beneath them, and floundering about in the dust of village streets.

The eggs of the giant water-bugs are attached to vertical stems of reeds just above the surface of the water. They are among the largest of insect eggs. Those of Benacus (fig. 118) are curiously striped. The eggs of a smaller, related water-bug, *Zaitha* or *Belostoma*, are attached by the female to the broad back of the

Fig. 118. Eggs of Benacus, enlarged; the lower-most are in process of hatching.

male, and are carried by him during their incubation. The nymphs of this family, on escaping from the egg suddenly unroll and expand their flat bodies, and attain at once proportions that would seem impossible on looking at the egg (fig. 119).

Most finely adapted to life in the water are the water boatmen (fig. 109 on p. 201) and the back-swimmers,

which swim with great agility and are able to remain for
a considerable time beneath the surface of the water.
The eggs of these are attached beneath the water to any
solid support. Most grotesque in form are the water-scorpions (Nepidæ), that breathe through a long caudal respiratory tube. The eggs of these are inserted into soft plant tissues, with a pair of long processes on the end of each egg left protruding.

FIG. 119. A new-hatched Benacus, and
a detached egg.

At the shore-line we find the creeping water-bugs among matted roots in the edge of the water, with shore
bugs and toad bugs just out on land.

Nymphs and adults alike are distinguished from the
members of all other orders by the possession of a
jointed puncturing and sucking proboscis beneath the
head, directed backward between the fore legs.

Nymphs and adults are found in the water together
and are alike carnivorous. Being similar in form they
are readily recognized as the same animal in different
developmental stages.

The net-winged insects (*Neuroptera*) are mainly
terrestrial or arboreal. Two families only have aquatic
representatives, the Sialididæ and the Hemerobiidæ,
and these are so different, they are better considered
separately.

1. *Sialididæ*—These are the dobsons, the fish flies and the orl flies. The largest is Corydalis, the common dobson (fig. 120), whose larva is the well known "hellgrammite", that is widely used as bait for bass. It lives under stones in rapids. It is a "crawler" of forbidding appearance, two or three inches long when grown, having a stout, greenish black body, sprawling, hairy legs, and paired fleshy lateral processes at the sides of the abdomen. There is a minute tuft of soft white gills under the base of each lateral process. There is a pair of stout fleshy prolegs at the end of the abdomen, each one armed with a pair of grappling hooks. The larvæ of the fish-flies (*Chauliodes*) are similar in form, but smaller and lack the gill tufts under

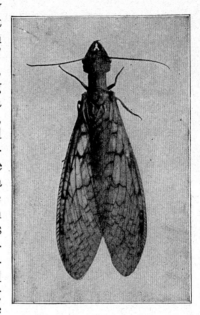

FIG. 120. An adult female dobson, *Corydalis cornuta*, natural size.

the lateral filaments. The larva of the orl-fly differs conspicuously in having no prolegs or hooks at the end of the body, but instead, a long tapering slender tail. Fish-fly larvæ are most commonly found clinging to submerged logs and timbers. Orl-fly larvæ burrow in the sandy beds of pools in streams and in lake shores.

All appear to be carnivorous, but little is known of the feeding habits of either larvæ or adults. Tho large and conspicuous insects they are rather secretive and are rarely abundant, and they have been little observed.

2. *Hemerobiidæ*—Of this large family of lace-wings but two small genera (in our fauna) of spongilla flies, *Climacia* and *Sisyra*, have aquatic larvæ. The adults are delicate little insects that are so secretive in habits

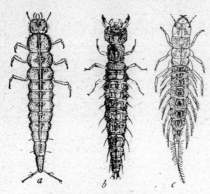

and so infrequently seen that they are rare in collections. Their larvæ are commonly found in the cavities of fresh water sponges. They feed upon the fluids in the body of the sponge. They are distinguished by the possession of long slender piercing mouthparts, longer than the head and thorax together, and by paired abdominal respiratory filaments, that are angled at the

FIG. 121. Insect larvae.

a, a diving-beetle larva (*Coptotomus interrogatus*), after Helen Williamson Lyman); *b*, a hellgrammite, (*Corydalis cornuta*, after Lintner); *c*, an orl-fly larva (*Sialis infumata*, after Maude H. Anthony).

base and bent underneath the abdomen. These larvæ are minute in size (6 mm. long when grown) and are quite unique among aquatic insect larvæ in form of mouthparts and in manner of life.

The caddis-flies (order *Trichoptera*) are all aquatic, save for a few species that live in mosses. They constitute the largest single group of predominantly aquatic insects. They abound in all fresh waters.

The adults are hairy moth-like insects that fly to lights at night, and that sit close by day, with their long antennæ extended forward (see fig. 103 on p. 197). They are not showy insects, yet many of them are very dainty and delicately colored. They are short-lived as adults, and, like the mayflies, many species swarm at the shore line on summer evenings in innumerable companies.

The larvæ of the caddis-flies mostly live in portable cases, which they drag about with them as they crawl or climb; but a few having cases of lighter construction, swim freely about in them. Such is *Triænodes*, whose spirally wound case made from bits of slender stems is shown in the accompanying figure.

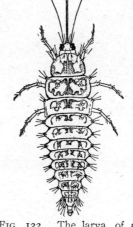

The cases are wonderful in their diversity of form, of materials and of construction. They are usually cylindric tubes, open at both ends, but they may be sharply quadrangular or triangular in cross section, and the tube may be curved or even coiled into a close spiral*.

FIG. 122. The larva of a spongilla fly, Sisyra (after Maude H. Anthony).

Almost any solid materials that may be available in the water in pieces of suitable size may be used in their case building: sticks, pebbles, sand-grains and shells are the staple materials. Sticks may be placed parallel and lengthwise, either irregularly, or in a continuous spiral. They may be placed crosswise with ends overlapping like the elements of a stick chimney, making thick walls and rather cumbrous cases.

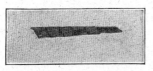

FIG. 123. The case of the free-swimming larvae of Triænodes.

However built, the case is always lined with the secretion from the silk glands of the larva. This substance is indeed the basis of all case construction. The larva

---

*As in Helicopsyche, (see fig. 221, on page 370) whose case of finely textured sand grains was originally described as a new species of snail shell.

builds by adding pieces one by one at the end of the tube, bedding each one in this secretion, which hardens on contact with the water and holds fast. Small snails and mussel shells are sometimes added to the exterior with striking ornamental effect, and sometimes these are added while the protesting molluscs are yet living in them.

FIG. 124. Cylindric sand cases of one of the Leptoceridae, (enlarged).

Some of the micro-caddis-flies (family Hydroptilidæ) fashion "parchment" cases of the silk secretion alone. These are brownish in color and translucent. They are usually compressed in form and are carried about on edge. *Agraylea* decorates the parchment with filaments of Spirogyra, arranged concentrically over the sides in a single external layer.

Some caddis-worms build no portable cases at all, but merely barricade themselves in the crevices between stones, attaching pebbles by means of their silk secretion, and thus building themselves a walled chamber which they line with silk. In this they live, and out of the door of the chamber they extend themselves half their length in foraging. Other caddisworms construct fixed tubes among the stones, and at the end of the tube that opens facing the current they spin fine-meshed funnel-shaped nets of silk. These are open up stream, and into them the current washes organisms suitable for food. The caddis-worm lies with ready jaws in wait at the bottom of the funnel, and cheerfully takes what heaven bestows, seizing any bit of food that may chance to fall into its net. These net-spinners belong to the family Hydropsychidæ.

When minute animals abound in the current the caddis-worms appear to eat them by preference: at other times, they eat diatoms and other algæ and plant fragments. The order as a whole tends to be herbivorous and many members of it are strictly so; but most of them will at least vary their diet with small may-fly and midge larvæ and entomostracans, when these are to be had.

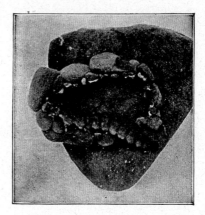

FIG. 125. The larva of *Rhyacophila fuscula* in its barricade of stones, exposed by lifting off a large top stone.

Caddis-worms are more or less caterpillar-like, but lack paired fleshy prolegs beneath the body, save for a single strongly-hooked pair at the posterior end. The thoracic legs are longer and stronger and better developed than in caterpillars, and they are closely applicable to the sides of the body, as befits slipping in and out of their cases. The front third of the body is strongly chitinized and often brightly pigmented; the remainder, that is constantly covered by the case, is thin skinned and pale. Most caddis-worms bear filamentous gills along the sides of the abdomen, but some that dwell in streams are gill-less and others have gills in great compound clusters or tufts.

Caddis-fly pupæ are likewise aquatic (and this is characteristic of no other order of insects), and like the larvæ, they often bear filamentous gills along the sides of the abdomen. They are equipped with huge mandibles that are supposed to be of use in cutting a way out through the silk just before transformation. The mandibles are shed at this time. The adult caddis-flies are destitute of jaws and are not known to feed; so they are probably short-lived.

FIG. 126. Eggs of Triænodes.

The eggs of caddis-flies usually are laid in clumps of gelatine. Sometimes they are arranged in a flat spiral, as in Triænodes, shown in the accompanying figure: sometimes they are suspended from twigs in a ring-like loop, as in Phryganea. Oftener they form an irregular clump. They are usually of a bright greenish color, but those of the net spinning Hydropsyches, laid on submerged stones in close patches with little gelatine, are tinged with a brick-red color.

The moths (order *Lepidoptera*) are nearly all terrestrial. Out of this great order of insects only a few members of one small family (Pyralidæ) have entered the water to live. These live as larvæ for the most part upon plants like water lilies and pond weeds that are not wholly submerged. *Hydrocampa*, removed from its case of two leaf fragments, looks like any related land caterpillar, with its small brown head, its strongly

FIG. 127. Two larval cases of the moth *Hydrocampa*, each made of two pieces of Marsilea leaf. Upper smaller case unopened, larva inside; lower case opened to show the larva, its cover below.

chitinized prothorax and the series of fleshy prolegs underneath the abdomen. By these same characters any other aquatic caterpillar may be distinguished from the members of other orders. *Paraponyx* makes no case, differs strikingly in being covered with an abundance of forking filamentous gills which surround the body as with a whitish fringe. It feeds, often in some numbers, on the under side of leaves of the white water-lily, or about the sheathing leaf bases of the broad-leaved pond weeds (*Potamogeton*).

Elophila fulicalis lives on· the exposed surfaces of stones in running streams, dwelling under a silt-covered canopy of thin-spun silk, about the edges of which it forages for algæ growing on the stones. Its body is

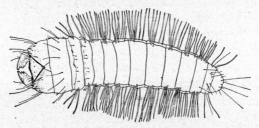

FIG. 128.   Larva of Elophila

depressed, and its gills are unbranched and in a double row along each side. It spins a dome-shaped cover having perforate margins under which to pass the pupal period. It emerges, to fly in companies of dainty little moths by the streamside.

All these aquatic caterpillars like their relatives on land, are herbivorous. They are all small species; they are of wide distribution and are often locally abundant.

The beetles (order *Coleoptera*) are mainly terrestrial, there being but half a dozen of the eighty-odd families of our fauna that are commonly found in the water. Both adults and larvæ are aquatic, but, unlike the bugs, the beetles undergo extensive metamorphosis, and

larvæ and adults are of very different appearance.
Beetle larvæ most resemble certain neuropteroids of the
family Sialididæ in appearance, and there is no single
character that will distinguish all of them (see fig. 121 on
p. 214). Only a few beetle larvæ (Gyrinids, and a few
Hydrophilids like Berosus) possess paired lateral filaments on the sides of the abdomen such as are characteristic of all the Sialididæ.

Aquatic beetle larvæ are
much like the larvæ of the
ground beetles (Carabidæ)
in general appearance, having well developed legs and
antennæ and stout rapacious
jaws.

Best known of water
beetles are doubtless the
"whirl-i-gigs" (Gyrinidæ),
which being social in their
habits and given to gyrating
in conspicuous companies on
the surface of still waters,
could hardly escape the
notice of the most casual observer. Their larvæ, how-
ever, are less familiar. They

FIG. 129. A diving beetle,
Dytiscus, slightly enlarged.

are pale whitish or yellowish translucent elongate creatures, with very long and slender paired lateral abdominal filaments along the sides of the abdomen.
They live amid the bottom trash where they feed upon
the body fluids of blood worms and other small
animal prey. Living often in broad expanses of shoal
water where there are no banks upon which to crawl
out for pupation, they construct a blackish cocoon
on the side of some vertical stem just above the surface
of the water and undergo transformation there. The

eggs are often laid on the under side of floating leaves of pondweeds.

Fig. 130.   One of the lesser diving beetles, *Hydro-porus*, seven times natural size.

The diving beetles (Dytiscidæ and Hydrophilidæ) are by far the most numerous and important of the aquatic beetles. These swarm in every pond and pool, and are among the most important carnivores of all such waters. They range in size from the big brown Dytiscus (fig. 129) down to little fellows a millimeter long. Their prevailing colors are brown or black, but many of the lesser forms are prettily flecked and

streaked with yellow (fig. 130). The eggs of the Dytiscus and of other members of its family are inserted singly into punctures in the tissues of living plants (fig. 131). Those of the Hydrophilids are for the most part inclosed in whitish silken cocoons attached to plants near the surface of the water.

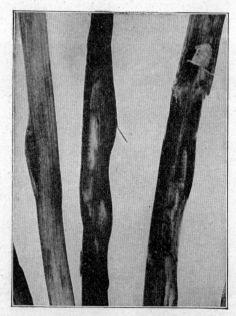

Fig. 131. Eggs of the diving beetle, *Dytiscus*, in submerged leafstalks, nearly ready for hatching: the larva shows through the shell. (From Matheson)

The Haliplids are a small family of minute beetles, having larvæ of unique form and habits. These larvæ

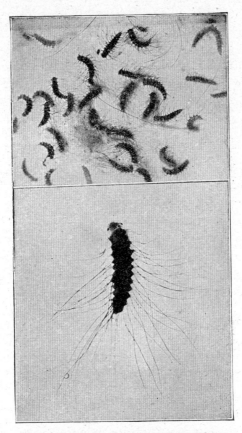

FIG. 132. Larvæ of the beetle, *Peltodytes*, in mixed algal filaments, twice natural size; below, a single larva more highly magnified. (From Matheson).

live among the tangled filaments of the coarser green algæ, especially Spirogyra, and they feed upon the contents of the cells that compose the filaments, sucking the contents of the cells, one by one. They are very inert-looking, stick-like, creatures and easily pass unobserved. Of our two common genera one (Peltodytes) is shown in figure 132. The body is covered over with very long stiff jointed bristle-like processes, giving it a burr-like appearance. The larva of the other genus (Haliplus) is more stick-like, has merely sharp tubercles upon the back, and has the body terminating in a long slender tail.

The Riffle beetles (Parnidæ and Amphizoidæ) prefer flowing water. They do not swim, but clamber over the surfaces of logs and stones. They are mostly small beetles of sprawling form, having stout legs that terminate in curved grappling claws. There is great variety of form among their larvæ, the better adapted ones that live in swift waters showing a marked tendency to assume a limpet-like contour. This culminates in the larva of Psephenus, commonly known as the "water penny." This larva was mistaken for a limpet by its original describer. It is very much flattened and broadened and nearly circular in outline, and the flaring lateral margins encircling and inclosing the body fit down all round to the surface of the stone on which it rests (see fig. 160 on page 260). Underneath its body are tufts of fine filamentous gills, intersegmentally arranged.

The flies (order *Diptera*) are a vast group of insects. Among them are many families whose larvæ are wholly or in part aquatic. The changes of form undergone during metamorphosis are at a maximum in this group: the larvæ are very different indeed from the adults.

Dipterous larvæ are very diversified in form and details of structure. The entire lack of thoracic legs will distinguish them from all other aquatic larvæ. They agree in little else than this, and the general tendency toward the reduction of the size of the head and of the appendages. Many of them are gill-less and many more possess but a single cluster of four tapering retractile anal gill filaments.

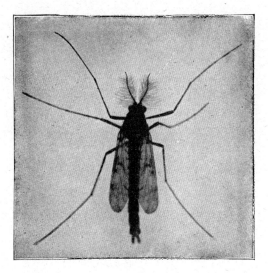

Fig. 133. An adult midge, *Tanypus carneus*, male.

By far the most important of the aquatic Diptera in the economy of nature are the midges (Chironomidæ). These abound in all fresh waters. The larvæ are cylindric and elongate, with distinct free head, and body mostly hairless save for caudal tufts of setæ. They are distinguished from other fly larvæ by the possession of a double fleshy proleg underneath the prothorax, and a pair of prolegs at the rear end of the body, all armed with numerous minute grappling hooks. Many of them are of a bright red color, and are hence called "blood worms."

Midge larvæ live mainly in tubes which they fashion out of bits of sediment held together by means of the secretion of their own silk glands. These tubes are built up out of the mud in the pond bottom as shown in the accompanying figure, or constructed in the crevices

FIG. 134.   Tubes of midge larvæ in the bed of a pool.

between leaves, or attached to stems or stones or any solid support. They are never portable cases. They are generally rather soft and flocculent. The pupal stage is usually passed within the same tubes and the pupa is equipped with respiratory horns or tufts of various sorts for getting its air supply. The pupa (see fig. 171 on p. 279) is active and its body is constantly undulating, as in the caddisflies.

The eggs of the midges are laid in gelatinous strings in clumps and are usually deposited at the surface of the water. Figure 135 shows the appearance of a bit of such an egg-mass. This one measured bushels in quantity, and doubtless was laid by thousands of midges. Figure 136 shows a little bit of it—a portion of a few egg strings—magnified so as to show the form and arrangement of the individual eggs. Such great egg masses are not uncommon, and they foreshadow the coming of larvæ in the water in almost unbelievable abundance.

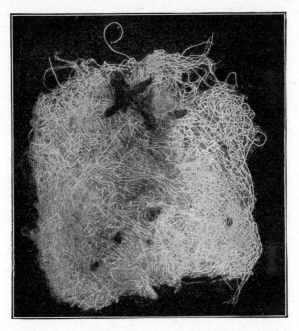

FIG. 135. A little bit of an egg mass of the midge, *Chironomus*, hung on water weeds (*Philotria*).

Midge larvæ are among the greatest producers of animal food. They are preyed upon extensively, and by all sorts of aquatic carnivores.

Three families of blood-sucking Diptera have aquatic larvæ; the mosquitoes (Culicidæ), the horseflies (Tabanidæ) and the black flies (Simuliidæ). Mosquito larvæ are the well known "wrigglers" that live in rain water barrels and in temporary pools. They are readily distinguished from other Dipterous larvæ by their swollen thoracic segments and their tail fin. The pupæ are free swimming and hang suspended at the surface with a pair of large respiratory horns or trumpets in contact with the surface when at rest.

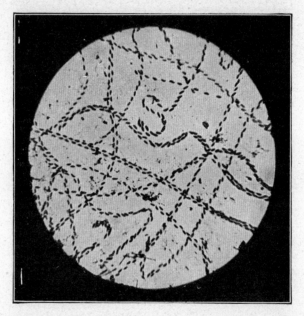

FIG. 136. A few of the component egg-strings, magnified.

The larvæ of the horseflies are burrowers in the mud of the bottom. They are cylindric in form, tapering to both ends, headless, appendageless, hairless, and have the translucent and very mobile body ringed with segmentally arranged tubercles. They are carnivorous, and feed upon the body fluids of snails and aquatic worms and other animals. The white spiny pupæ are

formed in the mud of the shore. The tiny black eggs
(fig. 138) are laid in close patches on the vertical stems
or leaves of emergent aquatic plants.

Black fly larvæ live in rapid streams, attached in
companies to the surfaces of rocks or timbers over
which the swiftest water pours. They are blackish,
and often conspicuous at a distance by reason of their
numbers. They have cylindric bodies that are swollen
toward the posterior end, which is attached to the
supporting surface by a sucking disc. Underneath the
mouth is a single median proleg, and on the front of the
head convenient to the mouth, there is a pair of "fans,"
whose function is to strain forage organisms out of the
passing current. The full grown larva spins a basket-
like cocoon on the vertical face of the rock or timber,
and in this passes its pupal stage. The eggs are laid
in irregular masses at the edge of the current where the
water runs swiftest.

In like situations we meet less frequently the net-
winged midges (Blepharoceridæ), whose scalloped flat
and somewhat limpet-shaped larvæ are at once recognizable by the possession of a midventral row of suckers
for holding on to the rock in the bed of the rushing
waters. The naked pupa is found in the same situation
and is attached by one strongly flattened side to the
supporting surface.

These five above-mentioned families are the ones
most given over to aquatic habits. Then there are
several large families a few of whose members are
aquatic: Leptidæ, whose larvæ live among the rocks
in rapid streams, hanging on and creeping by means of
a series of large paired and bifid prolegs; Syrphidæ,
whose larvæ are known as "rat-tailed maggots" since
their body ends in a long flexuous respiratory tube,
which is projected to the surface for air when the larva
lives in dirty pools; Craneflies (Tipulidæ) see fig. 215 on

FIG. 137. The larva of a horsefly, *Chrysops*.

FIG. 138. The eggs
of a horsefly on
an emergent bur-
reed leaf.

p. 360) whose cylindric tough-
skinned larvæ have their heads
retracted within the prothorax,
and bear on the end of the abdomen a respiratory disc perforate
by two big spiracles and surrounded by fleshy radiating filaments; minute moth-flies—Psycho-
didæ, (see fig. 214 on p. 359)
whose slender larvæ live amid
the trash in both brooks and
swales. Swaleflies (Sciomyzidæ)
whose headless and appendageless larvæ hang suspended by
their posterior end from the surface in still water; and others
less common.

It is a vast array of forms this
order comprises, this mighty group
of two-winged flies, that is still so
imperfectly known; and some of
the most highly diversified of its
larvæ are among the commoner
aquatic ones.

VERTEBRATES

There is little need that we should give any extended account of the groups of back-boned animals—fishes, amphibians, reptiles, birds and mammals. In water as on land they are the largest of animals, and are all familiar. The water-dwellers among them, excepting the fishes and a very few others, are air-breathing forms that are mainly descended from a terrestrial ancestry. They haunt the water-side and enter the shoals to forage or to escape enemies, but they cannot remain submerged, for they have need of air to breathe.

The fishes have remained strictly aquatic. They dominate the open waters of the larger lakes and streams. They have multiplied and differentiated and become adapted to every sort of situation where there is water of depth and permanence sufficient for their maintenance. They outnumber in species every other vertebrate group.

Within the water the worst enemies of fishes are other fishes; for the group is mainly carnivorous, and big fishes are given to eating little ones. Hence, tho all can swim, few of them do swim in the open waters, and these only when well grown. Those that so expose themselves must be fleet enough to escape enemies, or powerful enough to fight them. Little fishes and the greater number of mature fishes keep more or less closely to the shelter of shores and vegetation. The accompanying diagram, based on Hankinson's (08) studies at Walnut Lake, Michigan, represents the distribution of fishes in a rather simple case. The thirty-one species here present range in adult size from the pike which attains a length above three feet, to the least darter which reaches a length of scarcely an inch and a half. One species only, the whitefish,

dwells habitually in the deep waters of the lake. One
other species, the common sucker, is a regular inhabitant of water between fifteen and forty feet in depth.
The pike, ranges the upper waters at will pursuing his
prey over both depths and shoals; but he appears to
prefer to lie at rest among the water-weeds where his

FIG. 139. Ale-wives (*Clupea pseudoharengus*) on the beach of Cayuga Lake,
after the close of the spawning season. A single large sucker lies in the
foreground.

great mottled back becomes invisible among the lights
and shadows.

The pondweed zone on the sloping bottom between
five and twenty-five feet in depth is the haunt of most
of the remaining species, including all the minnows, catfishes, sunfishes, and the perches. The last named
wander betimes more freely into the deep water; all of

these forage in the shoals, especially at night. The catfishes are more strictly bottom feeders, and these feed mainly at night. A few species keep to the close shelter of thick vegetation at the water's edge, and one species, the least darter, prefers to lie over mottled marl-strewn bottoms at depth between fifteen and twenty feet.

So it appears that some two-thirds of the species have their center of abundance in the pondweed zone: here, doubtless they best find food and escape enemies.

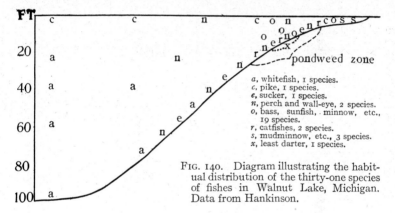

a, whitefish, 1 species.
c, pike, 1 species.
e, sucker, 1 species.
n, perch and wall-eye, 2 species.
o, bass, sunfish, minnow, etc.,
    19 species.
r, catfishes, 2 species.
s, mudminnow, etc., 3 species.
x, least darter, 1 species.

FIG. 140. Diagram illustrating the habitual distribution of the thirty-one species of fishes in Walnut Lake, Michigan. Data from Hankinson.

Only a few of the stronger and swifter species venture much into the deeper water: the weaklings and the little fishes frequent the weed-covered shoals.

The eggs of fishes are cared for in a great variety of ways. Their number is proportionate to the amount of nurture they receive. No species scatters its eggs throughout the whole of its range, but each species selects a spot more or less circumscribed in which to lay its young. Carp enter the shoals and scatter their eggs promiscuously over the submerged vegetation and the bottom mud with much tumult and splashing. A single female may lay upwards of 400,000 eggs a season.

Doubtless many of these eggs are smothered in mud and many others are eaten before hatching. Suckers seek out gravelly shoals, preferably in the beds of streams, at spawning time. Dangers are fewer here and a single female may lay 50,000 eggs. Yellow perch attach their eggs in strings of gelatin trailed over the surface of submerged water plants. The number per fish is still

Fig. 141. A splash on the surface made by a carp in spawning.

further reduced to some 20,000 eggs. Sunfishes make a sort of nest. They excavate for it by brushing away the mud with a sweeping movement of the pectoral fins. Thus they uncover the roots of aquatic plants over a circular area having a diameter equal to the length of the fish. On these roots the female lays her eggs, and the male guards them until they are hatched. With this additional care the number is further reduced to some 5000 eggs. Sticklebacks actually build a nest, by gathering and fastening together bits of vegetation. It is built in the tops of the weeds—not on the pond bottom. The nest is roughly spherical, with a hole through the middle of it from side to side. Within the dilated center of the passageway the female lays her eggs: the male stands guard over the nest. After the hatching of the eggs he still guards the young. It is said that when the young too early leave the nest, he catches them in his mouth and puts them back. The stickleback lays only about 250 eggs.

Thus in their extraordinary range of fecundity the fishes illustrate the wonderful balance in nature. For every species the number of young is sufficient to meet the losses to which the species is exposed.

The food of fresh-water fishes covers a very wide range of organic products; but the group as a whole is predaceous. A few, like the goldfishes and golden shiners, are mainly herbivorous and live on algæ and other soft plant stuffs. Others like carp and gizzard-shad live mainly on the organic stuffs they get by devouring the bottom ooze. Many, either from choice or from necessity, have a mixed diet of plant and animal foods. But the carnivorous habit is most widespread among them. In inland waters they are the greatest consumers of animal foods.

Such fishes as the pike which, when grown, lives wholly upon a diet of other fishes, are equipped with an abundance of sharp raptorial teeth. The sheepshead has flattened molar-like teeth strong enough for crushing shells and adapting it to a diet of molluscs. Other fishes, even large ones like the shovel-nosed sturgeon, have close-set gill-rakers. These retain for food the plancton organisms of the water that is strained through the gills. The young of all fishes are plancton feeders.

The Amphibians are the smallest of the five great groups of vertebrates. They are represented in our fauna mainly by frogs and salamanders. A few of the more primitive salamanders (Urodela), such as Necturus, breathe throughout life by means of gills, and are strictly aquatic. A few are terrestrial, but most are truly amphibious. They develop as aquatic larvæ (tadpoles), having gills for breathing and a fish-like circulation: they transform to air-breathing, more or less terrestrial adult forms; and they return to the water to lay their eggs in the primeval environment.

FIG. 142. A leopard frog. *Rana pipiens.*

The period of larval life varies from less than two months in the toad to more than two years in the bull-frog.

The eggs of amphibians are, for the most part, deposited in shallow water, often in masses in copious gelatinous envelopes (see fig. 201 on p. 342). In some cases the egg masses are large and conspicuous and well known. Examples are the long egg-strings of the toad that lie trailing across the weeds and the bottom; or the half-floating masses of innumerable eggs laid by the larger frogs. The eggs of the smaller frogs are less often seen, those of the peeper being attached singly to plant stems. Dr. A. H. Wright (14) has shown that the eggs of all our species of frogs are distinguishable by size, color, gelatinous envelopes and character of cluster.

Adult amphibians are carnivorous. They all eat lesser animals in great variety. Frogs and toads have a projectile and adhesive tongue which is of great service in capturing flying insects; but they eat, also, many other less active morsels of flesh that they find on the ground or in the water. The food of some of the lesser stream-inhabiting salamanders, such as Spelerpes, is mainly insects, while that of the vermilion-spotted newt is mainly molluscs.

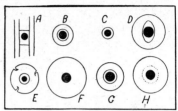

FIG. 143. Diagram of individual eggs from the egg mass of the toad and seven species of frog occurring at Ithaca. Eggs solid black; gelatinous envelopes white. (After Wright).

The amphibia are a group of very great biological interest. They represent a relatively simple type of vertebrate structure. Their development can be followed with ease and it is illuminating and suggestive of

A, Toad, eggs in double gelatinous tubes, forming strings, the inner tube divided by cross partitions; *B*, pickerel frog; *C*, peeper (no outer envelope); *D*, green frog (inner envelope ellipitical); *E*, tree frog (outer envelope ragged); *F*, bull frog (no inner envelope); *G*, leopard frog; *H*, wood frog. All twice natural size.

the early evolutionary history of the higher vertebrates. They illustrate in their own free-living forms the transition from aquatic to terrestrial life. And they show in the different amphibian types many grades of metamorphosis. The transformation is more extensive

FIG. 144. The spotted salamander, *Ambystoma tigrinum*.

in frogs than in any other vertebrates, involving profound changes in internal organs and in manner of life.

The reptiles are mainly terrestrial. Southward there are alligators in the water, but in our latitude there are

Fig. 145. The common snapping turtle.

only a few turtles and water snakes. These make their nests on land. They hide their eggs in the sand or in the midst of marshland rubbish, where the sun's warmth incubates them.

These also are carnivorous.

The water birds, tho more numerous than the two preceding groups, are but a handful of this great class of vertebrates.

The principal kinds of birds that frequent the water are water-fowl—ducks, geese and swans; the shore birds—plover, snipe and rails; the gulls, the herons and the divers. Some of these that, like the loon, are

FIG. 146. Wild geese foraging in a marsh in Dakota.

superably fitted for swimming and diving, feed mainly on fishes. Most water birds consume a great variety of lesser animals. The ducks and rails differ much in diet according to species. Thus the Sora rail eats mainly seeds of marsh plants, while the allied Virginia rail in the same locality eats miscellaneous animal food to the extent of more than fifty per cent. of its diet.

Only the waterfowl that are prized as game birds are extensively herbivorous. They eat impartially the vegetable products of the land and of the water. The wild ducks and geese eat great quantities of duckmeat (Lemna) and succulent submerged aquatics. Canvas-backs fatten on the wild celery (Vallisneria). In Cayuga Lake in winter they gorge themselves with the starch-filled winter buds of the pondweed, *Potamogeton*

Fig. 147. Floating nest of pied-billed grebe (*Podilymbus podiceps*) in a cat-tail marsh, surrounded by water.

pusillus. They also dive and pluck up from the bottom mud the reproductive tubers of the pondweed, *Potamogeton pectinatus* (see fig. 228 on p. 381).

Water birds, having attained the freedom of the air, are wide ranging beyond all other animals. They come and go in annual migrations. They settle here and there, and commit local and intermittent depredations. The water birds nest mainly on land, and in their nesting and brooding habits they differ little from their terrestrial relatives.

The aquatic mammals of inland waters fall mainly in two groups, the carnivores and the rodents. Here again, the carnivores that are more expert swimmers and divers, such as fisher, martin, otter and mink are all fish-eating animals. They have become fitted to

FIG. 148. A muskrat, *Fiber zibethicus.*

utilize the chief animal product of the water. Of these four the mink alone has withstood the "march of progress," and retains its former wide distribution.

Of rodents there are two fur-bearers of much import-ance, the beaver, now driven to the far frontier, and the muskrat. The muskrat has become under modern agricultural conditions the most important aquatic mam-mal remaining. By reason of its rapid rate of repro-duction, its ability to find a living in any cat-tail marsh, big or little, and its hardiness, it has been able to main-tain its place.

CHAPTER V

ADJUSTMENT TO CONDITIONS OF AQUATIC LIFE

INDIVIDUAL ADJUSTMENT

SO infinitely varied are the fitnesses of aquatic organisms for the conditions they have to meet that we can only select out of a worldful of examples a few of the more widespread and significant. We shall have space here for discussing only such adaptations to life in the water as are common to large groups of organisms, and represent general modes of adjustment. First we will consider some of the ways in which the species is fitted to the aquatic conditions under which it lives, and then we will take note of some mutual adjustments between different species.

The first of living things to appear upon the earth were doubtless simple organisms that were far from being so small as the smallest now existing, or so large as the largest. They grew and multiplied. They differentiated into plants and animals, into large and small, into free-swimming and sedentary. Some betook themselves to the free life of the open waters and others to more settled habitations on shores. The open-water forms were nomads, forever adrift in the waves: the shoreward forms might find shelter and a quiet resting place.

LIFE IN OPEN WATER

In the open water there are certain great advantages that lie in minuteness and in buoyancy. These qualities determine the ability of organisms to float freely about in the more productive upper strata of water. To descend into the depths is to perish for want of light. So the members of many groups are adapted for floating and drifting about near the surface. These constitute the *plancton*.

On the other hand, large size has its advantages when coupled with good ability for swimming and food gathering. In the rough world's strife the battle is usually to the strong. It is the larger, wide-ranging, free-swimming organisms that dominate the life of the open water. These constitute the *necton*.

Plancton and necton will be discussed in the next chapter as ecological groups, but in this place we may take note of the two very different sorts of fitness, that they have severally developed for life in the open water, the plancton organisms being fitted for flotation, and the necton for swimming.

Flotation—All living substance is somewhat heavier than water (i. e. has a specific gravity greater than 1) and therefore tends to sink to the bottom. The velocity in sinking is determined by several factors, one of which is external and the others are internal:

The external factor is the varying viscosity of the water.

The internal factors are specific gravity, form and size.

We have mentioned (p. 30) that the viscosity of the water is twice as great at the freezing point as at ordinary summer temperatures; which means, of course, that the water itself would offer much greater resistance to the sinking of a body immersed in it. We are here concerned with the internal factors.

Lessening of specific gravity—The bodies of organisms are not composed of living substance alone, but contain besides, inclusions and metabolic products of various sorts, which oftentimes alter their specific gravity. The shells and bone and other hard parts of animals are usually heavier than protoplasm; the fats and gelatinous products and gases are lighter. We know that the fats of vertebrates, if isolated and thrown upon the water, will float; and that a fat man, in order to maintain himself above the water, needs put forth less effort than a lean one. There are probably many products of the living body that are retained within or about it and that lessen its specific gravity, but the commonest and most important of these seem to fall into three groups:

1. *Fats and oils*, which are stored assimilation products. These are very easily seen in such plancton organisms as Cyclops (see fig. 96 on p. 189) where they show through the transparent shell as shining yellowish oil droplets. Most plancton algæ store their reserve food products as oils rather than as starches.

2. *Gases*, which are by-products of assimilation, and are distributed in bubbles scattered through the tissue

where produced, or accumulate in special containers. These greatly reduce the specific gravity of the body, enabling even heavy shelled forms (see p. 159) to float.

3. *Gelatinous and mucilaginous* products of the body which usually form external envelopes (see fig. 10 on p. 52) but which may appear as watery swellings of the tissues. Their occurrence as envelopes is very common with plants and with the eggs of aquatic animals; they may serve also for protection and defense, and for regulating osmotic pressure, but by reason of their low specific gravity they also serve for flotation.

Improvment of form—We have already called attention (p. 42) to the fact that size has much to do with the rate of sinking in still water. This is because the resistance of the water comes from surface friction and the smaller the body the greater the ratio of its surface to its mass. Given a body small enough, its mere minuteness will insure that it will float. But in bodies of larger size relative increase in surface is brought about in various ways:

1. By extension of the cell in slender prolongations (see fig. 50, j, k, l, on p. 129).
2. By the aggregation of cells into expanded colonies:
    a. Discoid colonies, as in Pediastrum (fig. 44 on p. 123).
    b. Filaments, as in Oscillatoria (fig. 34 on p. 109).
    c. Flat ribbons of innumerable slender cells placed side by side, as in many lake diatoms (Fragillaria, Tabelaria, Diatoma).
    d. Radiate colonies as in Asterionella (fig. 35 *n* on p. 111).
    e. Spherical colonies as in Volvox (fig. 31, p. 105: see also *a b c* of fig. 50 on p. 129), wherein the cells are peripheral and widely separated the

interstices and the interior being filled with gelatinous substances of low specific gravity.

f.   Dendritic colonies, as in Dinobryon (fig. 32 on p. 106).

3.   In the Metazoa, by the expansion of the external armor and appendages into bristles, spines and fringes. Thus in the rotifer *Notholca longispina* (fig. 149), a habitant of the open water of lakes, there is a great prolongation of the angles of the lorica, before and behind; and in the Copepods (fig. 95, p. 188) there is an extensive development of bristles upon antennæ and caudal appendages.

Expansions of the body, if mere expansions, serve only to keep the body passively afloat; but many of them have acquired mobility, becoming locomotor organs.   Cilia and flagella are the simplest of these, and are common to plants and animals.   Almost all the appendages of the higher animals, antennæ, legs, tails, etc., are here and there adapted for swimming.   A body whose specific gravity is but little greater than that of the water may be sustained by a minimum use of swimming apparatus.   The lesser

FIG. 149.
A long-
spined
rotifer.

flagellate and ciliate forms, both plant and animal, maintain their place by continuous lashing of the water.   If we watch a few waterfleas in a breaker of clear water we shall see that their swimming also, is unceasing.   Each one swims a few strokes of the long antennæ upward, and then settles with bristles all outspread, descending slowly, as resistance yields, to its former level.   This it repeats again and again.   It may turn to right or to left, rise a little higher or sink a little lower betimes, but it keeps in the main to its proper level.   Its swimming powers are to an important degree supplemental to its inadequate powers of flotation. The strokes of its swimming antennæ are, like the beating of our own hearts, intermittent but unceasing, and when these fail it falls to its grave on the lake bottom.

Flotation devices usually impede free swimming, especially do such expansions of the body as greatly increase surface contact with the water. It is in the resting stages of animals, therefore, that we find the best development of floats: such, for example, as the overwintering statoblasts of the Bryozoan, Pectinatella, shown in the accompanying figure. Here an encysted mass of living but inactive cells is surrounded by a buoyant, air-filled annular cushion, as with a life preserver, and floats freely upon the surface of the water, and is driven about by the waves.

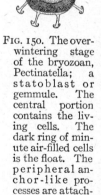

FIG. 150. The overwintering stage of the bryozoan, Pectinatella; a statoblast or gemmule. The central portion contains the living cells. The dark ring of minute air-filled cells is the float. The peripheral anchor-like processes are attachment hooks for securing distribution by animals.

Too great buoyancy is, however, as much a peril to the active micro-organisms of the water as too little. Contact with the air at the surface brings to soft protoplasmic bodies, the peril of evaporation. Entanglement in the surface film is virtual imprisonment to certain of the water-fleas, as we shall see in the next chapter. It is desirable that they should live not on but near the surface. A specific gravity about that of water would seem to be the optimum for organisms that drift passively about: a little greater than that of water for those that sustain themselves in part by swimming.

Terrestrial creatures like ourselves, who live on the bottom in a sea of air with solid ground beneath our feet, have at first some difficulty in realizing the nicety of the adjustment that keeps a whole population in the water afloat near to, but not at the surface. This comes out most clearly, perhaps, in those minor changes of form that accompany seasonal changes in temperature of the water. In summer when the viscosity of the water grows less (and when in consequence its resist-

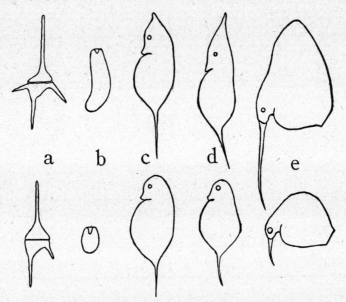

FIG. 151. Summer and winter forms of plancton animals: summer above, winter below. *a*, the flagellate Ceratium; *b*, the rotifer Asplanchna; *c*, *d*, *e*, water-fleas; *c* and *d*, Daphne; *e*, Bosmina. (After Wesenberg-Lund).

ance to sinking is diminished) the surface of many plancton organisms is increased to correspond. The slender diatoms grow longer and slenderer, the spines on certain loricate rotifers grow longer. Bristles and hairs extend and plumes and fringes grow denser. Even the form of the body is altered to increase surface-contact with the water. A few examples are shown in the accompanying figures. These changes when followed thro the year show a rather distinct correspondence to the seasonal changes in viscosity of the water.

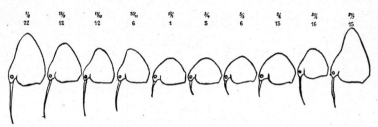

Fig. 152. Seasonal form changes of the water-flea, *Bosmina coregoni*. The fractional figures above indicate date: those below indicate corresponding temperatures in °C. (After Wesenberg-Lund.)

Swimming—For rapid locomotion through the water there are numberless devices for propulsion, but there is only one thoroly successful form of body; and that

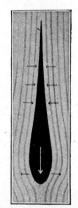

Fig. 153. Stream-line form. For explanation see text.

is the so-called "stream-line form" (fig. 153). It is the form of body of a fish: an elongate tapering form, narrowed toward either end, but sloping more gently to the rear. It is also the form of body of a bird encased in its feathers. It is probably the form of body best adapted for traversing any fluid medium with a minimum expenditure of energy. The accompanying diagram explains its efficiency. The white arrow indicates direction of movement. The gray lines indicate the displacement and replacement of the water. The black arrows indicate the direction in which the forces act. At the front the force of the body is exerted against the water; at the rear the force of the water is exerted against the body. The water, being perfectly mobile, returns after displacement; and much of the force expended in pushing it aside at the front is regained by the return-push of the water against the sloping rearward portion of the body.

The advantage of stream-line form is equally great whether a body be moving through still water, or whether it be standing against moving water. A mackerel swimming in the sea is benefited no more than is a darter holding its stationary position on the stream bed. To this we shall have occasion to return when discussing the rapid-water societies.

Apparatus for propulsion is endlessly varied in the different animal groups. Plants have developed hardly any sort of swimming apparatus beyond cilia and flagella. These also serve the needs of many of the lower animals—the protozoa, the flat worms, the rotifers, trochophores and other larvæ, sperm cells generally, etc. But more widely ranging animals of larger size have developed better swimming apparatus, either with or without appendages. Snakes swim by means of horizontal undulating or sculling movements of the body, and so also do many of the common minute Oligochæte worms. Horseleeches swim in much the same manner, save that the undulations of the body are in the vertical plane. Midge larvæ ("bloodworms") swim with figure-of-8-shaped loopings of the body that are quite characteristic. Mosquito larvae are "wrigglers," and so also are many fly and beetle larvæ, tho each kind wriggles after its own fashion. Dragonfly nymphs swim by sudden ejection of water from the rectal respiratory chamber.

All of these swim without the aid of movable appendages; but the larger animals swim by means of special swimming organs, fringed and flattened in form and having an oar-like function. These may be fins, or legs, or antennæ, or gill plates, in infinite variety of length, form, position and design.

Great is the diversity in aspect and in action of the animals that swim. Yet it is perfectly clear, even on a casual inspection, that the best swimmers of them all are those that combine proper form of body—stream-line form—with caudal propulsion by means of a strong tail-fin.

LIFE ON THE BOTTOM

Shoreward, the earth beneath the waters gives aquatic organisms an opportunity to find a resting place, a temporary shelter, or a permanent home. Flotation devices and ability at swimming may yet be of advantage to the more free-ranging forms; but the existence of possible shelter and of solid support makes for a line of adaptations of an entirely different sort. Here dwell the aquatic organisms that have acquired heavy armor for defense; heavy shells, as in the mussels; heavy carapaces as in the crustaceans; heavy chitinous armor as in the insects; or heavy incrustations of lime as in the stoneworts.

The condition of the bottom varies from soft ooze in still water to bare rocks on wave washed shores. The differences are very great, and they entail significant differences in the structure of corresponding plant and animal associations. These have been little studied hitherto, but a few of the more obvious adaptations to bottom conditions may be pointed out in passing.

First we will note some adaptations for avoidance of smothering in silt on soft bottoms; then some adaptations for finding shelter by burrowing in sandy bottoms and by building artificial defenses: then some adaptations for withstanding the wash of the current on hard bottoms.

I

Avoidance of silt—Gills are essentially thin-walled expansions of the body, that provide increased surface for contact with the water, and thus promote that exchange of gases which we call respiration. Gills usually develop on the outside of the body; for it is only in contact with the water that they can serve their function. In most animals that live in clear waters they are freely exposed upon the outside; but in animals that live on soft muddy bottoms they are withdrawn into protected chambers (or, rather, sheltered by the outgrowth of surrounding parts) and fresh water is passed to them thro strainers. Thus the gills of a crawfish occupy capacious gill chambers at the sides of the thorax, and water is admitted to them thro a set of marginal strainers. The gills of fresh-water mussels are located at the rear of the foot within the inclosure of valves and mantle, and water is passed to and from them thro the siphons. The gills of dragonfly nymphs are located on the inner walls of a rectal respiratory chamber, and water to cover them is slowly drawn in thro a complicated strainer that guards the anal aperture, and then suddenly expelled thro the same opening, the valves swinging freely outward.

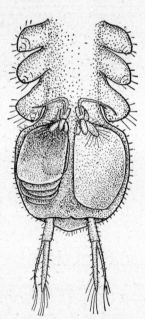

FIG 154. The abdomen of Asellus, inverted, showing gill packets.

There is probably no better illustration of parallel adaptation for silt avoidance than that furnished by the

crustacean, Asellus, and the nymph of the mayfly,
Cænis. Both live in muddy bottoms where there is
much fine silt. Both possess paired plate-like gills.
In Asellus they are developed underneath the abdomen;
in Cænis upon the back. In Asellus they are double;
in Cænis, simple. In
Asellus they are blood
gills; in Cænis, tracheal
gills. In both they are
developed externally in
series, a pair corresponding to a body segment.
In both they are soft and
white and very delicate.
But in both an anterior
pair has been developed
to form a pair of enlarged
opercula or gill covers.
These are concave pos-
teriorly and overlie and
protect the true gills.
The gills have been ap-
proximated more closely,
so that they are the more
readily covered over; and
they have developed in-
terlacing fringes of radiating marginal hairs,
which act as strainers,
when the covers are raised to open the respiratory
chamber.

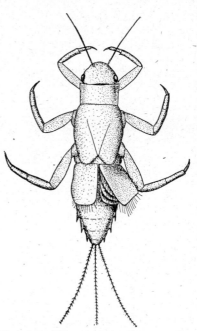

Fig. 155. The nymph of the mayfly
Cænis, showing dorsal gill packets.

Such are the mechanical means whereby suffocation
in the mud is avoided. It must not be overlooked that
there is a physiological adaptation to the same end. A
number of soft bodied thin-skinned animals have an
unusual amount of hæmoglobin in the blood plasma—

enough, indeed, to give them a bright red color. This substance has a great capacity for gathering up oxygen where the supply is scanty, and of yielding it over to the tissues as needed. True worms that burrow in deep mud, and Tubifex (see fig. 83 on p. 174) that burrows less deeply and the larger bright red tube making larvæ of midges known as "blood worms" (see fig. 236 on p. 393) are examples. Since these forms live in the softest bottoms, where the supply of oxygen is poorest, where few other forms are able to endure the conditions, their way of getting on must be of considerable efficiency.

II

Burrowing—The ground beneath the water offers protection to any creature that can enter it; protection from observation to a bottom sprawler, that lies littered over with fallen silt; protection from attack about in proportion to its hardness, to anything that can burrow.

Animals differ much in their burrowing habits and in the depth to which they penetrate the bottom. Many mussels and snails burrow very shallowly, pushing their way along beneath the surface, the soft foot covered, the hard shell-armored back exposed. The nymphs of Gomphine dragonflies (fig. 116 on p. 209) burrow along beneath the bottom with only the tip of the abdomen exposed at the surface of the mud. Other insect larvæ descend more deeply into burrows which remain open to the water above: while horsefly larvæ and certain worms descend deeply into soft mud.

The two principal methods by which animals open passageways thro the bottom are (1) by digging, and (2) by squeezing thro. Digging is the method most familiar to us, it being commonly used by terrestrial animals. Squeezing thro is the commonest method of aquatic burrowers.

FIG. 156. A nymph of a burrowing mayfly, Ephemera. (From *Annals Entom. Soc. of America:* drawing by Anna H. Morgan).

The digging of burrows requires special tools for moving the earth aside. These, as with land animals, are usually flattened and shovel-like fore legs. The other legs are closely appressed to the body to accommodate them to the narrow burrow. The hind legs are directed backward. The head is usually flattened and more or less wedge-shaped, and often specially adapted for lifting up the soil preparatory to advancing thro it (see fig. 116 on p. 209).

One of the best exponents of the burrowing habit is the nymph of the may-fly, Hexagenia, whose innumerable tunnels penetrate the beds of all our larger lakes and rivers. It is an ungainly creature when exposed in open water; but when given a bed of sand to dig in, it shows its fitness. Be-

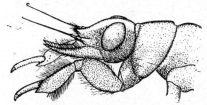

FIG. 157. The front of a burrowing mayfly nymph, Hexagenia, much enlarged, showing the pointed head, the great mandibular tusks and the flattened fore legs.

sides having feet that are admirably fitted for scooping the earth aside, it has a pair of enormous mandibular tusks projecting forward from beneath the head. It thrusts forward its approximated blade-like fore feet, and with them scrapes the sand aside, making a hole. Then it thrusts its tusks into the bottom of the hole and lifts the earth forward and upward. Then, moving forward into the opening thus begun and repeating these operations, it quickly descends from view.

Squeezing thro the bottom is the method of progress most available to soft-bodied animals. Those lacking hard parts such as shovels and tusks with which to dig make progress by pushing a slender front into a narrow opening, and then distending and, by blood pressure enlarging the passageway. The horsefly larva shown in figure 137 on page 230 (discussed on page 227) is a good example. The body is somewhat spindle-shaped, taper-ing both ways, and adapted for traveling forward or backward. It is exceedingly changeable in proportions being adjustable in length, breadth and thickness. Indeed, the whole interior is a moving mass of soft organs, any one of which may be seen thro the trans-parent skin, slipping backward or forward inside for a distance of several segments. The body wall is lined with strong muscles inside, and outside it bears rings of stout tubercles, which may be drawn in for passing, or set out rigidly to hold against the walls of the burrow. The extraordinary adjustability of both exterior and interior is the key to its efficiency. When such a larva wishes to push forward in the soil, it distends and sets its tubercles in the rear* to hold against the walls, and drives the pointed head forward full length into the mud; then it compresses the rear portion, forcing the blood

---

*Certain cranefly larvæ (such as *Pedicia albivitta* and *Eriocera spinosa*) that live in beds of gravel have one segment near the end of the body expansible to almost balloon-like proportions, forming a veritable pushing-ring in the rear.

forward to distend the body there, thus widening the burrow. And if anyone would see how such a larva gets through a narrow space when the walls cannot be pushed farther apart, let him wet his hand and close the larva in its palm; the larva will quickly slip out between the fingers of the tightly closed hand; and when half way out it will present a strikingly dumb-bell-shaped outline. Here, again, we see the advantage of its almost fluid interior.

This adjustability of body, is of course, not peculiar to soft bodied insect larvæ; it is seen in leeches and slugs and many worms.

The mussel's mode of burrowing is not essentially different from that above described. The slender hollow foot is pushed forward into the sand, and then distended by blood forced into it from the rear. When sufficiently distended to hold securely by pressure against the sand, a strong pull drags the heavy shell forward.

III

Shelter building—Some animals produce adhesive secretions that harden on contact with the water. Thus, these are able to bind loose objects together into shelters more suitable for their residence than any that nature furnishes ready made. The habit of shelter building has sprung up in many groups; in such protozoans as Difflugia (see fig. 69 *c* on p. 39); in such worms as Dero (see fig. 82 on p. 174); in such rotifers as Melicerta (see fig. 86 on p. 178); in such caterpillars as Hydrocampa (see fig. 127 on p. 219); in nearly all midges, as Chironomus (see figure 134 on p. 226) and Tanytarsus (see fig. 223 on p. 373); and especially in the caddis-worms, all of which construct shelters of some sort and most of which build portable cases. The extraordinary prevalence in all fresh waters of such forms as the larvæ of midges and caddis-flies would indicate that the habit has been biologically profitable.

According to Betten the habit probably began with the gathering and fastening together of fragments for a fixed shelter, and the portable, artifically constructed, silk lined tubes of the higher caddis-worms are a more recent evolution.

IV

Withstanding the wash of moving waters—Where waters rush swiftly, mud and sand and all loose shelters

FIG. 158. Stone from a brook bed, bearing tubes of midge larvæ and portable cases of two species of caddis-worms. The more numerous spindle-shaped cases are those of the micro-caddisworms of the genus *Hydroptila*. For more distinct midge tubes see figs. 134 and 223.

are swept away. Only hard bare surfaces remain, and the creature that finds there a place of residence must build its own shelter, or must possess more than ordinary advantages for maintaining its place. The gifts of the gods to those that live in such places are chiefly these three:

1. Ability to construct flood-proof shelters. Such are the fixed cases of the caddis-worms and midge larvæ (fig. 158) to which we shall give further consideration in the next chapter.

2. Special organs for hanging on to water-swept surfaces. Such organs are the huge grappling claws of the nymphs of the larger stoneflies (see fig. III on p. 204) and of the riffle beetles: also powerful adhesive suckers, such as those of the larvæ of the net-winged midges.

FIG. 159. The larva of the net-winged midge, Blepharocera, dorsal and ventral views.

3. Form of body that diminishes resistance to flow of the water. This we have already seen is stream-line form. In our discussion of swimming we pointed out that the form of body that offers least resistance to the progress of the body through the water will also offer least resistance to the flow of water past the body. So we find the animals that stand still in running water are of stream-line form; darters and other fishes of the rapids; mayflies, such as Siphlonurus and Chirotenetes; even such odd forms as the larvæ of Simulium, which hangs by a single sucker suspended head downwards in the stream. Indeed, the case of Simulium is especially significant, for with the reversal of the position of the body the greater widening of the body is shifted from the anterior to the posterior end, and stream-line form is preserved. Such forms as these live in the open, remain for the most part quietly in one position and wait for the current to bring their food to them.

There are other more numerous forms living in rapid water that cling closer to the solid surfaces, move about upon and forage freely on these surfaces, and the adaptations of these are related to the surfaces as much

as to the open stream. These have to meet and withstand the water also, but only on one side; and the form is half of that of our diagram (fig. 153). It is that figure divided in the median vertical plane, with the flat side then applied to the supporting surface, and flattened out a bit at the edges. This is not fish form, but it is the form of a limpet. This is the form taken on by a

FIG. 160.   Limpet-shaped animals. At right the larva of the Parnid beetle, Psephenus, known as the "water-penny." At left, the snail, Ancylus.

majority of the animals living in rapid waters. When the legs are larger they fall outside of the figure, as in the mayfly shown on page 369, and are flattened and laid down close against the surface so as to present only their thin edges to the water. When the legs are small, as in the water-penny, (fig. 160) they are covered in underneath. Sometimes there are no legs, as in the flatworms, and in the snail, Ancylus.

Here, surely, we have the impress of environment. Many living beings of different structural types are moulded to a common form to meet a common need; and even the non-living shelters built by other animals are fashioned to the same form. The case of the micro-caddisworm, *Ithytrichia confusa* (fig. 161) is also limpet-shaped;

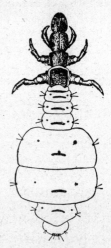

FIG. 161.   The larva of the caddis-worm, *Ithytrichia confusa*.

so also is the pupal shelter of the caterpillar of *Elophila fulicalis;* hardly less so is the portable case of the larva of the caddis-fly, *Leptocerus ancylus* or of *Molanna angustata.*

ADJUSTMENT OF THE LIFE CYCLE

Life runs on serenely in the depths of the seas where, as we have noted in Chapter II, there is no change of season; but in shoal and impermanent waters it meets with great vicissitudes. Winter's freezing and summer's drouth, exhaustion of food and exclusion of light and of air, impose hard conditions here. Yet in these shoals is found perhaps the world's greatest density of popula-

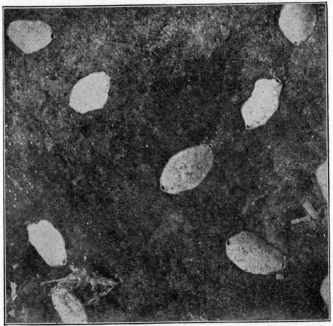

FIG. 162. The flattened and limpet-shaped cases of Ithytrichia confusa, as they appear attached to the surface of a submerged stone.

tion.   Here competition for food and standing room is
most severe.   And here are made some of the most
remarkable shifts for maintaining "a place in the sun."

Encystment—The shifts which we are here to consider
are those made in avoidance of the struggle—shifts
which have to do with the tiding over of unfavorable
seasons by withdrawal from activity.   This means
encystment or encasement of some sort or in some
degree.   The living substance secretes about itself
some sort of a protective layer, and, enclosed within it,
ceases from all its ordinary functions.

This is the most familiar to us in the reproductive
bodies of plants and animals; in the zygospores of
Spirogyra and desmids and other conjugates; in the
fruiting bodies of the stoneworts; in the seeds of the
higher plants; and in the over-wintering eggs of many
animals.   Most remarkable perhaps is the brief seasonal
activity of forms that inhabit temporary pools.   Such
Branchipods as Chirocephalus (see fig. 90 on p. 184)
Estheria and Apus, appear in early spring in pools
formed from melting snow.   They run a brief course of
a few weeks of activity, lay their eggs and disappear to
be seen no more until the snows melt again.   Their
eggs being resistant to both drying and freezing, are
able to await the return of favorable conditions for
growth.   The eggs of Estheria have been placed in
water and hatched after being kept dry for nine years.

But it is not alone reproductive bodies that thus tide
over unfavorable periods.   The flatworm, *Planaria
velata*, divides itself into pieces which encyst in a layer
of slime and thus await the return of conditions favorable for growth.   The copepod, *Cyclops bicuspidatus*,
according to Birge and Juday (09) spends the summer
in a sort of cocoon composed of mud and other bottom
materials rather firmly cemented together about its

body. It forms this cocoon about the latter end of
May. It reposes quietly upon the bottom during the
entire summer—thro a longer period, indeed, than
that of absence of oxygen from the water. Hatching and resumption of activity begin in September and
continue into October. Marsh (09) suggests that
with us this species "may be considered preëminently a

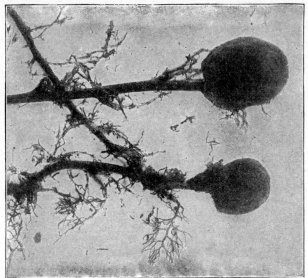

FIG. 163. Hibernacula of the common bladderwort.

winter form." It is active in summer only in cold
mountain lakes.

The over-wintering buds (hibernacula) of some aquatic seed plants are among the simplest of these devices.
Those of the common bladderwort are shown in figure
163. At the approach of cold weather the bladderwort
ceases to unfold new leaves, but develops at the tip of
each branch a dense bud composed of close-laid incompletely developed leaves. This is the hibernaculum.
It is really an abbreviated and undeveloped branch.

Unlike other parts of the plant, its specific gravity is greater than that of water. It is enveloped only by a thin gelatinous covering. With its development the functional activity of the old plant ceases; the leaves lose chlorophyl; their bladders fall away; the tissues

FIG. 164. The remains of a fresh-water sponge that has grown upon a spray of water-weed. The numerous rounded seed-like bodies embedded in the disintegrating tissue are statoblasts. See text.

disintegrate; and finally the hibernacula fall to the bottom to pass the winter at rest. When the water begins to be warmer in spring, the buds resume growth, the axis lengthens, the leaves expand, air spaces develop and gases fill them, buoying the young shoots up into better light, and the activities of another season are begun.

Statoblasts—Perhaps the most specialized of over-wintering bodies are those of the Bryozoans and fresh-water sponges, known as statoblasts. These are little masses of living cells invested with a tough and hard and highly resistent outer coat. They are formed within the flesh of the parent animal (as indicated for Bryozoan in fig. 77 on p. 167), and are liberated at its dissolution (as indicated for a sponge in the accompanying figure). They alone survive the winter. As noted earlier in this chapter, their chitinous coats are often expanded with air cavities to form efficient floats: sometimes in Bryozoan statoblasts there is added to this a series of hooks for securing distribution by animals (see fig. 150 on p. 247). Often in autumn at the Cornell Biological Field Station collecting nets become clogged with these hooked statoblasts.

In the fresh-water sponges the walls of the statoblast are stiffened with delicate and beautiful siliceous spicules, and there is at one side a pore through which the living cells find exit at the proper season. Since marine sponges lack statoblasts, and some fresh-water species do not have them, it is probable that they are an adaptation of the life cycle to conditions imposed by shoal and impermanent waters.

Winter Eggs—Another seasonal modification of the life cycle is seen in the Rotifers and water-fleas. Here there are produced two kinds of eggs; summer eggs that develop quickly and winter eggs that hibernate. The summer eggs for a long period produce females only. They develop without fertilization. In both these groups males are of very infrequent occurrence. They appear at the end of the season. The last of the line of parthenogenetic females produce eggs from which hatch both males and females and the last crop of eggs is fertilized. These are the over-wintering eggs.

The accompanying figures illustrate both kinds of eggs in the water-flea, Ceriodaphnia, an inhabitant of bottomland ponds. Figure 165 shows a female with the summer eggs in the brood chamber on her back. These thin-shelled eggs are greenish in color. They hatch where they are and the young Ceriodaphnias live

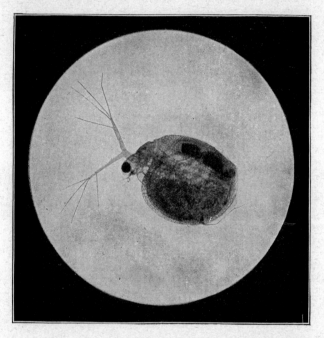

Fig. 165.   Ceriodaphnia, with summer eggs.

within the brood-chamber until they have absorbed all the yolk stored within the egg and have become very active. Then they escape between the valves of the shell at the rear.

Winter eggs in this species are produced singly. Figure 166 shows one in the brood chamber of another female. It is inclosed in a chitinized protective covering, which, because of its saddle-shaped outline, is called an *ephippium*. This egg is liberated unhatched by the molting of the female, as shown in figure 167. It remains in its ephippium over winter, protected from freezing, from drouth and from mechanical injury,

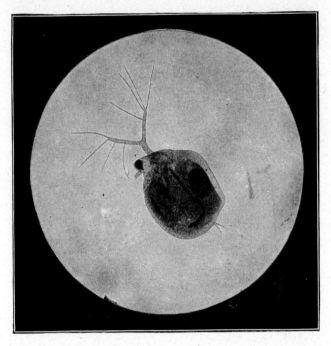

FIG. 166. Ceriodaphnia bearing an ephippium containing the single winter egg.

and buoyed up just enough to prevent deep submergence in the mud of the bottom. With the return of warmer weather it may hatch and start a new line of parthenogenetic female Ceriodaphnias.

Thus, it is that many organisms are removed from our waters during a considerable part of the winter season.

The water-fleas and many of our rotifers are hibernating as winter eggs.   The bryozoans and sponges are hibernating as statoblasts.   Doubtless many of the simpler organisms whose ways are still unknown to us have their

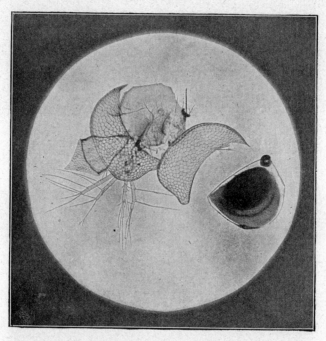

FIG. 167.   Ceriodaphnia, molted skin and liberated ephippium of the same individual shown in the preceding figure.   This photograph was taken only a few minutes after the other. The female after molting immediately swam away.

own times and seasons and modes of passing a period of rest.   It is doubtless due, also, to the ease and safety with which they may be transported when in such condition that they all have a wide distribution over the face of the earth.   In range, they are cosmopolitan.

Readaptations to life in the water—The more primitive groups of aquatic organisms have, doubtless, always been aquatic; but the aquatic members of several of the higher groups give evidence of terrestrial ancestry. Among the reasons for believing them to have developed from forms that once lived on land is the possession of characters that could have developed only under terrestrial conditions, such as the stomates for intake of air in the aquatic vascular plants, the lungs of aquatic mammals, and the tracheæ and spiracles of aquatic insects. Furthermore, they are but a few members (relatively speaking) of large groups that remain predominantly terrestrial in habits, and there are among them many diverse forms, fitted for aquatic life in very different ways, and showing many signs of independent adaptation.

I

The vascular plants are restricted in their distribution to shores and to shoal waters. They are fitted for growth in fixed position and they possess a high degree of internal organization with a development of vessels and supporting structures that cannot withstand the beating of heavy waves. As compared with the land plants of the same groups, these are their chief structural characteristics:

1. In root:—reduced development. With submergence there is less need of roots for food-gathering, since absorption may take place over the entire surface. Roots of aquatic plants serve mainly as anchors; in a few floating plants as balancers; sometime they are entirely absent.

2. In stems:—many characteristics, chief of which are the following:

    a. Reduction of water-carrying tubes, for the obvious reason that water is everywhere available *b.* Reduction of wood vessels and of wood fibers and other mechanical tissues. In the denser medium of the water these are not needed, as they are in the air, to support the body. Pliancy, not rigidity, is required in the water.

    c. Enlargement of air spaces. This is prevalent and most striking. One may grasp a handful of any aquatic stems beneath the water and squeeze a cloud of bubbles out of them.

    d. Concentration of vessels near the center of the stem where they are least liable to injury by bending.

    e. A general tendency toward slenderness and pliancy in manner of growth, brought about usually by elongation of the internodes.

  3. In leaves:—many adaptive characters; among them these:

    a. Thinness of epidermis, with absence of cuticle and of ordinary epidermal hairs. This favors absorption through the general surfaces.

    b. Reduction of stomates, which can no longer serve for intake of air.

    c. Development of chlorophyl in the epidermis, which, losing the characters which fit it for control of evaporation, takes on an assimilatory function.

    d. Isolateral development, i. e., lack of differentiation between the two surfaces.

    e. Absence of petioles.

    f. Alteration of leaf form with two general tendencies manifest: Those growing in the most stagnant waters become much dissected (bladderworts, milfoils, hornworts, crowfoots, etc.).

    Those growing in the more open and turbulent waters become long, ribbonlike, and **very** flexible (eelgrass, etc.).

4. In general, the following characteristics:
   a. The production of abundance of mucilage, which, forming a coating over the surface, may be of use to the plants in various ways:
     1. For flotation, when the mucilage is of low specific gravity.
     2. For defense against animals to which the mucilage is inedible or repugnant.
     3. For lubrication: a very important need; for, when crossed plant stems are tossed by waves, the mucilage reduces their mutual friction and prevents breaking.
     4. For preventing evaporation on chance exposure to the air.
     5. For regulating osmotic pressure, and aiding in the physical processes of metabolism.
   b. Development of vegetative reproductive bodies:
     1. Hibernacula, such as those of the bladderwort (fig. 162).
     2. Tubers such as those of the sago pondweed (see fig. 228), the arrow-head, etc.
     3. Burs, such as terminate the leafy shoots of the ruffled pondweed (see fig. 63).
     4. Offsets and runners, such as are common among land plants.
     5. Detachable branches and stem segments, that freely produce adventitious roots and establish new plants.
   c. Diminished seed production. This is correlated with the preceding. Some aquatics such as duckweeds and hornworts are rarely known to produce seeds; others ripen seeds, but rarely develop plants from them. Their increase is by means of the vegetative propagative structures above mentioned, and they hold their place in the world by continuous occupation of it.

II

The mammals that live in the water are two small orders of whales, Cetacea and Sirenia, and a few scattering representatives of half a dozen other orders. Tho few in number they represent almost the entire range of mammalian structure. They vary in their degree of fitness for water life from the shore-haunting water-vole, that has not even webbing between its toes, to the ocean going whales, of distinctly fish-like form, that are entirely seaworthy. It is a fine series of adaptations they present.

For all land-animals, returned to the water to live, there are two principal problems, (1) the problem of getting air and (2) the problem of locomotion in the denser medium. Warm-blooded animals have also the problem of maintaining the heat of the body in contact with the water. To begin with the point last named, aquatic mammals have solved the problem of heat insulation by developing a copious layer of fat and oils underneath the skin. This development culminates in the extraordinary accumulation of blubber in arctic whales.

No aquatic mammals have developed gills. They all breathe by means of lungs as did their terrestrial ancestors. All must come to the surface for air. Their respiratory adaptations are slight, consisting in the shifting of the nostrils to a more dorsal position and providing them with closable flaps or valves, to prevent ingress of the water during submergence.

It is with reference to aquatic locomotion that mammals show the most striking adaptations. About in proportion to their fitness for life in the water they approximate to the fish-like contour of body that we have already discussed (page 249) as stream-like form. Solidity and compactness of the anterior portion of the body are brought about by consolidation of the neck vertebræ and shortening of the cranium. Smoothness of contour, (and therefore diminished resistance to passage through the water) is promoted by (1) the loss of hair; (2) the loss of the external ears; (3) the shortening and deflection of the basal joints of the legs; (4) elongation of the rear portion of the body. Caudal propulsion is attained in the whales by the huge dorsally flattened tail; in the seals (whose ancestors were perhaps tailless) by the backwardly directed hind legs.

Compared with these marine mammals those of our fresh waters show very moderate departures from terrestrial form. The beaver has broadly webbed hind feet for swimming. The muskrat has a laterally flattened tail. The mink, the otter and the fisher, with their elongate bodies and paddle-like legs, are best fitted for life in the water, and spend much time in it. But all fresh-water mammals make nests and rear their young on land.

III

The insects that live in the water have adaptations for swimming that parallel those of mammals, just noted; but some other adaptations grow out of the different nature of their respiratory system, and, more grow out of the difference in their life cycle. The free-living larval stage of insects offers opportunity for independent adaptation in that stage. Adult insects of but two orders, Coleoptera and Hemiptera, are commonly found in the water. These, as compared with their terrestrial relatives, exhibit many of the same adaptations already noted in mammals; (1) approximation to stream-line form, with (2) consolidation of the forward parts of the body for greater rigidity; (3) lowering of the eyes and smoothing of all contours; (4) loss of hair and sculpturing, and (5) shortening of basal segments of swimming legs, with lengthening of their oar-like tips, flattening and flexing of them into the horizontal plane, and limiting their range of motion to horizontal strokes in line with the axis of gravity of the body. Caudal propulsion does not occur with adult insects; none of them has a flexible tail. Oar-like hind feet are the organs of propulsion. The best swimmers among them are a few of the larger beetles: Cybister, which swims like a frog with synchronous strokes of its powerful hind legs, and Hydrophilus, with equally good swimming legs, which, like the whale, has developed a keel for keeping its body to rights.

Adult insects, like the mammals, lack gills, and rise to the surface of the water for air; but they take the air not through single pairs of nostrils, but a number of pairs of spiracles, and they receive it, not into lungs, but into tracheal tubes that ramify throughout the body. The spiracles are located at the sides of the thorax and abdomen, in general a pair to each segment.

In diving beetles the more important of these are the ones located on the abdomen beneath the wings. Access to these is between the wing tips. The beetles when taking air hang at the surface head downward. The horny, highly arched, fore wings are fitted closely to the body to inclose a capacious air chamber. They are opened a little at their tips for taking in a fresh air supply at the surface. Then they are closed, and the beetle, swimming down below, carries a store of air with him.

In other beetles there are different methods of gathering and carrying the air. The little yellow-necked beetles of the famiiy Haliplidæ, gather the air with the fringed hind feet, pass it forward underneath the huge ventral plates which, in these beetles cover the bases of the hind legs, and thence it goes through a transverse groove-like passage (fig. 168) to a chamber underneath the wing bases, where there are two enlarged spiracles on each side. The beetles of the family Hydrophilidæ have their ventral surface covered with a layer of fine water-repellant pubescence, to which the air readily adheres. Thus the air is carried exposed upon the surface, where it shines like a breastplate of silver.

In the waterbugs, the air is usually carried on the back under the wings, but the inverted back-swimmers conduct air to their spiracles through longitudinal

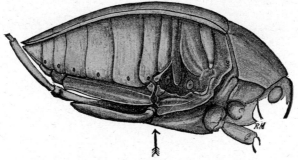

FIG. 168. Diagram of the air-taking apparatus of the beetle, Haliplus. The arrow indicates the transverse groove that leads to the air chamber. (From Matheson.)

grooves that are covered by water-repellant hairs, and that extend forward from the tip of the abdomen upon the ventral side. The water walking-stick, Ranatra, and some of its allies have developed a long respiratory tube out of a pair of approximated grooved caudal stylets. This long tail-like tube reaches the surface while the bug stays down below, breathing like a man in a diving bell.

The immature stages of aquatic insects are far more completely adapted to life in the water than are the adults. Some members of nearly all the orders, and all the members of a few of the smaller orders live and grow up in the water.   These facts have been noted, group by group, in Chapter IV.   Here we may explain that the reason for this probably lies in the greater plasticity of the immature stages.   All are thin-skinned on hatching from the egg, and a supply of oxygen may be taken from the water by direct absorption thro the general surface of the body.   With growth gills develop; but these have no relation to the structure or life of the adult and are lost at the final transformation.

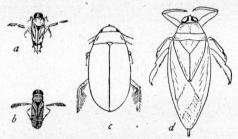

FIG. 169.   Adult aquatic insects: *a*, the back swimmer (Notonecta); *b*, the waterboatman (Corixa); *c*, a diving beetle (Dytiscus); *d*, a giant water-bug (Benacus).

Here again we find all degrees of adaptation.   The larvæ of the long-horned leaf beetles (Donacia, etc.) that live wholly submerged have solved the problem of getting air by attaching themselves to plants and perforating the walls of their internal air spaces, thus tapping an adequate and dependable air supply that is rich in oxygen.   This method is followed also by the larvæ of several flies and at least one mosquito.   There are many aquatic larvæ that breathe air at the surface as do adult bugs and beetles.   Some of these, such as the swaleflies and craneflies, (fig. 215) differ little from their terrestrial relatives.   Others like the mosquito are specialized for swimming and breathe thro respiratory trumpets.   A few like the rat-tailed maggot parallel the method of Ranatra mentioned above in that they have developed a long respiratory tube, capable of reaching the surface of the water while they remain far below.

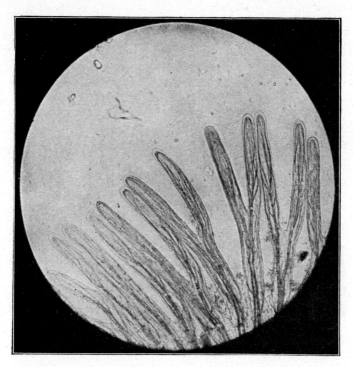

FIG. 170. Tracheal gill of the mayfly nymph, Heptagenia, showing loops of tracheoles toward the tip.

Of those that breathe the air that is dissolved in the water a few lack gills even when grown to full size; but these for the most part live in well aerated waters, and possess a copious development of tracheæ in the thinner portions of their integument. Such are the pale nymphs of the stonefly, Chloroperla, that live in the rapids of streams and the slender larvæ of the punkie Ceratopogon, that live where algæ abound.

The gills of insect larvæ are of two principal sorts: blood-gills and tracheal gills. Blood-gills are cylindric outgrowths of the integument into which the blood flows. Exchange of gases is between the blood inside the gills and the water outside. Such gills are most commonly appended to the rear end of the alimentary canal, a tuft of four retractile anal gills being common to many dipterous larvæ. Bloodworms have also two pairs developed upon the outside wall of the penultimate segment of the body (see fig. 236 on p. 393). Such gills are most like those of vertebrates.

Tracheal gills are more common among insect larvæ. These are similar outgrowths of the skin, traversed by fine tracheal air-tubes. In these the exchange of gases is between the water and the air contained within the tubes, and distribution of it is thro the complex system of tracheæ that ramify throughout the body. The tracheæ where they enter such a gill usually split up into long fine multitudinous tracheoles that form recurrent loops, rejoining the tracheal branches (fig. 170).

Tracheal gills differ remarkably in form, position and arrangement. In form they are usually either slender cylindric filaments, or small flat plates. Filamentous gills are more common, only this sort occurring on stone-fly nymphs (fig. 111 on p. 204), and on caddis-worms. Lamelliform or plate-like gills occur on the back of may-flies (fig. 113), and on the tail of damselflies (fig. 115). Either kind may grow singly or in clusters. Filamentous gills are often branched. In the stonefly, Tæniop-teryx, they are unbranched but composed of three somewhat telescopic segments. Both filamentous and lamelliform gills occur on many mayflies.

There is another form of tracheal gills, sometimes called "tube gills" developed upon the thorax of many dipterous pupæ. Whatever their form they are merely hollow bare chitinous prolongations from the mouth of the prothoracic spiracle. They are expanded "respiratory trumpets" in mosquito pupæ, branching horns in black-fly pupæ, and fine brushes of silvery luster in bloodworm pupæ. No pupæ, save those of the caddisflies, have tracheal gills of the ordinary sort.

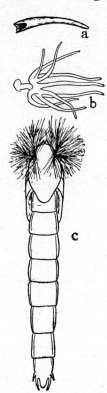

FIG. 171. Tube-gills of Dipterous pupæ: *a*, of a mosquito, Culex; *b*, of a blackfly, Simulium; *c*, of a midge, Chironomus. (*a* and *b* detached).

Gills are developed rarely on the head, more often on the thorax, and very frequently on the abdomen. They grow about the base of the maxillæ in a few stonefly and mayfly nymphs, about the bases of the legs in most stonefly nymphs and almost anywhere about the sides or end of the abdomen in all the groups. They are ventral in the spongilla flies, dorsal in the mayflies, lateral in the orl-fly and beetle larvæ, caudal in the damselflies, anal in most dipterous larvæ, and they cover the inner walls of a rectal respiratory chamber in dragonflies. Such extraordinary diversity in structures that are so clearly adaptive is perhaps the strongest evidence of the independent adaptation of many insect larvæ to aquatic life.

Propulsion by means of fringed swimming legs occurs in a few insect larvæ, such as the caddis-worm, Triænodes, and the "water-tiger" Dytiscus. The gill plates of many mayflies and damselflies are provided with muscles, and these are used for swimming. Caudal propulsion is also the rule in these same groups. Among beetle and fly larvæ locomotion is mainly effected by wrigglings of the body, that are highly individualized but only moderately efficient, if judged by speed.

It is worthy of note that the completest adaptations to conditions of aquatic life do not occur in those groups of insects that are aquatic in both adult and larval stages. Beetle larvæ and water-bug nymphs take air at the surface, and in structure differ but little from their terrestrial relatives. Fine developments of tracheal gills occur in the nymphs of mayflies and stoneflies, and in caddis worms; internal gill chambers, in the dragonfly nymphs; attachment apparatus for withstanding currents, in some dipterous larvæ; the utmost adaptability to all sorts of freshwater situations occurs in the midges; and in adult life these insects are all aerial.

What then is the explanation of the dominance of this remarkable insect group in the world to-day—a dominance as noteworthy in all shoal freshwaters as it is on land? What advantages has this group over other groups? There is no single thing; but there are two things that, taken together, may give the key to the explanation. These are:

1. *Metamorphosis*, the changes of form usually permitting an entire change of habitat and of habits between larval and adult life. The breaking up of the life cycle into distinct periods of growth and reproduction permits development where food abounds.

2. *The power of flight* in the adult stage permits easy getting about for finding scattered sources of food supply and for laying eggs.

In quickly growing animals no larger than insects these matters are very important; for even a small and transient food supply may serve for the nurture of a brood of larvæ.   And if the food supply be exhausted in one place, or if other conditions fail there, the adults may fly elsewhere to lay their eggs.   The facts of dominance would seem to justify this explanation, since those groups that most abound in the world to-day are in general the ones in which metamorphosis is most complete and in which the power of flight is best developed.

II. MUTUAL ADJUSTMENT

VARIOUS phenomena of association between non-competing species are manifest alike in terrestrial and aquatic societies. The occurrence of producers and consumers is universal. Carnivores eat herbivores, and parasites and scavengers follow both in every natural society. Symbiosis is as well illustrated in green hydra and green ciliates as in the lichens. The mutually beneficial association between fungus and the roots of green plants is as well seen in the bog as in the forest. The larger organisms everywhere give shelter to the smaller, and many examples, such as that of the alga, Nostoc, that dwells in the thallus of Azolla, or the rotifer *Notommata parasita* that lives in the hollow internal cavity of Volvox, occur in the water world.

We shall content ourselves here with a very brief account of two associations, one of which has to do mainly with a mode of getting a living, the other with providing for posterity. The first will be insectivorous plants; the second the relations between fishes and fresh-water mussels.

I

Insectivorous plants—The plants that capture insects and other animals for food are a few bog plants such as sundew and pitcher-plant, and a number of submerged bladderworts. These have turned tables on the animal world. Living where nitrogenous plant-foods of the ordinary sorts are scanty, they have evolved ways of availing themselves of the rich stores of proteins found in the bodies of animals. The sundew seems to digest its prey like a carnivore; the bladderwort absorbs the dissolved substance like a scavenger. Charles Darwin studied these plants fifty years ago, and his account ('75) is still the best we have.

The sundew, Drosera, captures insects by means of an adhesive secretion from the tips of large glandular hairs

FIG. 172. A leaf of sundew with a captured caddis-fly. The glandular hairs are bent downward, their tips in contact with the body of the insect. Other erect hairs show globules of secretion enveloping their tips.

that cover the upper surface of its leaves (fig. 172). The leaves are few in number and spatulate in form, and are laid down in a rosette about the base of a stem, flat upon the mud or upon the bed of mosses in the midst of which Drosera usually grows. They are red in color, and crowned and fringed with these purple hairs, each with a pearly drop of secretion at its tip sparkling in the light, like dew, they are very attractive to look upon.   The insect that makes the mistake of settling upon one of these leaves is held fast by the tips of the hairs it touches: the more it struggles the more hairs it touches, and the more firmly it is held.   Ere it ceases its struggles all the hairs within reach of it begin to bend over toward it and to apply their tips to the surface of its body.   Thus it becomes enveloped with a host of glands, which then pour out a digestive secretion upon it to dissolve its tissues.   When digested its substance is absorbed into the tissue of the leaf.

The pitcher-plant, Sarracenia, captures insects in a different way.   Its leaves are aquatic pitfalls.   They rise usually from the surface of the sphagnum in a bog (see fig. 207 on p. 350) on stout bases from a deep seated root stalk.   They are veritable pitchers, swollen in the middle, narrowed at the neck and with flaring lips. The rains fill them.   Insects fall into them and are unable to get out again; for all around the inner walls in the region of the neck there grows a dense barrier of long sharp spines with points directed downward. This prevents climbing out.   The insects are drowned, and their decomposed remains are absorbed by the plant as food.

It is mainly aerial insects that are destroyed, flies, moths, beetles, etc.; and we should not omit to note in passing that there are other insects, habituated to life in the water of the pitchers, and that normally develop there.   Such are the larvæ of the mosquito, *Aedes smithi*, and of a few flies and moths.

The bladderworts (Utricularia) are submerged plants that float just beneath the surface.   On their bright green, finely dissected leaves are innumerable minute traps (not bladders or floats as the name of the plant implies) having the appearance shown in the accompanying figure. These capture small aquatic animals, such as insect larvæ, crustaceans, mites, worms, etc.

The mechanism of the trap is shown diagrammatically in figure 174. First of all there is a circle of radiating hairs about the entrance, set diagonally outward, like the leaders of a fisherman's fyke net, and well adapted to turn the free-swimming water-flea toward the

Fig. 173. A spray of the common bladderwort, Utricularia.

proper point of ingress. Then there is a transparent elastic valve stopping the entrance, hinged by one side so that it will readily push inward, but holding tightly against the rim when pressed outward. This

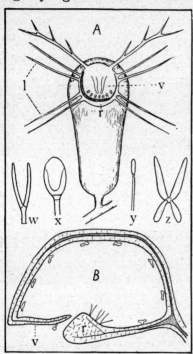

is the most important single feature of the trap. It makes possible getting in easily and impossible getting out at all. Darwin speaks of a Daphnia which inserted an antennæ into the slit, and was held fast during a whole day, being unable to withdraw it. On the outer face of the valve near its margin is a row of glandular hairs. These have roundly swollen terminal secreting cells. They may be alluring in function, tho this has not been proven. Directed backward across the center of the valve are four stiff bristles, that may be useful for keeping out of the passageway animals too big to pass through it—such as might blockade the entrance.

FIG. 174. Diagram of the mechanism of a trap of one of the common bladderworts. A, The trap from the ventral side, showing the outspread leader hairs converging to the entrance, l. leaders, r. rim, v, valve. B, A median section of the same r, rim; v, valve; w, x, y, z, epidermal hairs; w, from the inner side of the rim; x, from the free edge of the valve; y, from the base of the valve; z, from the general inner surface of the trap.

Small animals when entrapped swim about for a long time inside, but in the end they die and are decomposed. New traps are of a bright translucent greenish color; old ones are blackish from the animal remains they contain. The inner surface of the trap is almost completely covered with branched hairs. These are erect forked hairs adjacent to the rim, and flat-topped four-rayed hairs over the remainder of the wall space. These hairs project into the dissolved fluids, as do roots into the nutriet solutions in the soil, and their function is doubtless the absorption of food.

II

The larval habits of fresh-water mussels—The early life of our commonest fresh-water mussels is filled with

FIG. 175. Small minnows bearing larval
mussels (glochidia) on their fins.

shifts for a living that illustrate in a remarkable way the interdependence of organisms. The adult mussels burrow shallowly through the mud, sand and gravel of the bottom (as noted on page 108) or lie in the shelter of stones. Their eggs are very numerous, and hatch into minute and very helpless larvæ. For them the vicissitudes of life on the bottom are very great. The chief peril is perhaps that of being buried alive and smothered in the mud. In avoidance of this and as means of livelihood during early development the young of mussels have mostly taken on parasitic habits. They attach themselves to the fins and gills of fishes (fig. 175). There they feed and grow for a season, and there they undergo a metamorphosis to the adult form. Then they fall to the pond bottom and thereafter lead independent lives.

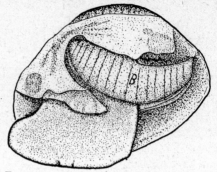

FIG. 176.   A gravid mussel (*Symphynota complanata*) with left valve of shell and mantle removed, showing brood pouch (modified gill) at B. (After Lefevre and Curtis.)

The eggs of the river-mussels are passed into the watertubes of the gills where they are incubated for a time. Packed into these passageways in enormous numbers they distend them like cushions, filling them out in various parts of one or both gills according to the species, but mostly filling the outer gill. When one picks up a gravid mussel from the river bed the difference between the thin normal gill and the gill that is serving as a brood chamber (fig. 176) is very marked.

Glochidia—In the case of a very few river mussels (*Anodonta imbecillis*, etc.) development to the adult form occurs within the brood chamber; but in most river mussels the eggs develop there into a larval form that is known as a *glochidium*. This is already a bivalve (fig. 177) possessing but a single adductor muscle for closing the valves and lacking the well developed system of nutritive organs of the adult. It is very sensitive to contact on the ventral surface. In this condition it is cast forth from the brood chamber.

If now the soft filament of a fish's gill, or the projecting ray of a fin by any chance comes in contact with this sensitive surface the glochidium will close upon it almost with a snap; and if the fish be the right kind for the fostering of this particular mollusc, it will remain attached. It is indeed interesting to see how manifestly ready for this reaction are these larvæ. If a ripe brood chamber of Anodonta (fig. 88 on p. 180) be emptied into a watch glass of water, the glochidia scattered over the bottom will lie gaping widely and will snap their toothed valves together betimes, whether touched or not. And they will tightly clasp a hair drawn across them.

Doubtless gills become infected when water containing the glochidia is drawn in through the mouth and passed out over them. Fins by their lashing cause in the water swirling currents that bring the glochidia up against their soft rays and thin edges.

Glochidia vary considerably in form and size, in so much that with careful work species of mussels can usually be recognized by the glochidia alone. Thus it is possible on finding them attached to fishes, to name the species by which the fishes are infected.

In size glochidia range usually between .5 and .05 millimeter in greatest diameter. Some are more or less triangular in lateral outline and these have usually a pair of opposed teeth at the ventral angle of the valves. Others are ax-head shaped and have either two teeth or none at all on the ventral angles. But the forms that have the ventral margin broadly rounded and toothless are more numerous. Whether toothed or not they are able to cling securely when attached in proper place to a proper host.

The part taken by the fish in the association is truly remarkable. The fish is not a mere passive agent of mussel distribution. Its tissues repond to the stimulus

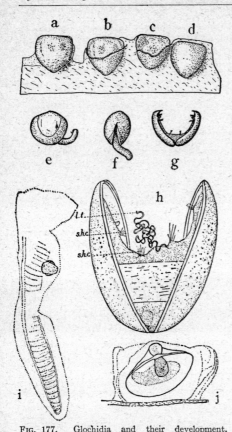

of the glochidia in a way that parallels the response of a plant to the stimulus of a gall insect. As a plant develops a gall by new growth of tissue about the attacking insect, and shuts it in and both shelters and feeds it, so the fish develops a cyst about the glochidium and protects and feeds it. The tissues injured by the valves of the glochidium produce new cells by proliferation. They rise up about the larva and shut it in (fig. 177). They supply food to it until the metamorphosis is complete, and then, when it is a complete mussel in form, equipped with a foot for burrowing and with a good system of nutritive organs, they break away from it and allow it to fall to the bottom. Since this period lasts for some weeks, or even in a few cases, months, the fishes by

Fig. 177. Glochidia and their development. into larval mussels, *a, b, c, d,* stages in the encystment of glochidia of the mussel, Anodonta, on the fin of a carp; *e* and *f,* young mussels (Lampsilis) a week after liberation from the fish; *g,* glochidium of the mussel, Lampsilis, before attachment. (After Lefevre and Curtis).

h, glochidium of the wash-board mussel, *Quadrula heros,* greatly enlarged and stained to show the larval thread (*l t*) and sensory hair cells (*s h c*) The clear band is the single adductor muscle.

i, a gill filament of a channel cat-fish bearing an encysted glochidium of the warty-back mussel: the cyst is set off by incisions of the filament. The darker areas on the edges of the valves indicate new growth of mussel shell. (After Howard.)

j, Encysted young of *Plagiola donaciformis,* showing great growth of adult shell, beyond the margin of glochidial shell—much greater growth than occurs in most species during encystment. (After Surber.)

wandering from place to place aid the distribution of the mussels, but they do much more than this.

It is to be noted, furthermore, that this relation is a close one between particular species, just as it is between plants and gall insects. Each attacking species has its own particular host. Recent careful studies made by Dr. A. D. Howard and others at the Fairport Biological Laboratory have shown such relations as the following:

Species of Mussels		*Host Species*
1. Yellow Sand Shell	(Lampsilis anodontoides)	on the gars
2. Lake Mucket	(Lampsilis luteolus)	on the basses and perches
3. Butterfly Shell	(Plagiola securis)	on the sheepshead
4. Warty Back	(Quadrula pustulosa)	on the channel catfish
5. Nigger-head	(Quadrula ebeneus)	on the blue herring
6. Missouri Nigger-head	(Obovaria ellipsis)	on the sturgeons
7. Salamander mussel	(Hemilastena ambigua)	on Necturus

Some of these mussels infect one species of fish; some, the fishes of one family or genus; a few have a still wider range of host species, these last being usually the species having the larger and stronger glochidia with the best development of clasping hooks on the valve tips. A very special case is that of Hemilastena, a mussel that lives under flat stones and projecting rock ledges in the stream bed. Living in the haunts of the mud-puppy, Necturus, and out of the way of the fishes, it infects the gills of this salamander with its glochidia.

The glochidia will grow only on their proper hosts. They will take hold on almost any fish that touches them in a manner to call forth their snapping reaction, but they will subsequently fall off from all but their proper hosts, without undergoing development.

Whether it be the mussel that reacts only to a certain kind of fish substance, or the fish that reacts to form a cyst only for a certain glochidial stimulus is not known. The relation appears onesided, and beneficial only to the parasitic mussel; yet moderate infestation appears to do little harm to the fishes. The cysts are soon grown, emptied and sloughed off, leaving no scar. And a few fishes, such as the sheepshead which is host for many mussels, appear to reap an indirect return, in that their food consists mainly of these same mussels when well grown.

It may be noted in passing that one little European fish, the bitterling, has turned tables on the mussels. It possesses a long ovipositor by means of which it inserts its own eggs into the gill cavity of a mussel, where they are incubated.

CHAPTER VI

AQUATIC SOCIETIES

LIMNETIC SOCIETIES

GREAT bodies of water furnish opportunity for all the different lines of adaptation discussed in the preceding chapter. The sun shines full upon them in all its life-giving power. The rivers carry into them the dissolved food substances from the land. Wind and waves and convection currents distribute these substances throughout their waters. Both the energy and the food needed for the maintenance of life are everywhere present. Here are expanses of open water for such organisms as can float or swim. Here are shores for such as must find shelter and resting places; shores bare and rocky; shores low and sandy; shores sheltered and muddy, with bordering marshes and with inflowing streams. The character of the population in any place is determined primarily by the fitness of the organisms for the conditions they have to meet in it.

Aquatic Societies

For every species the possible range is determined by climate; the possible habitat, by distribution of water and land; the actual habitat, by the presence of available food and shelter, and by competitors and enemies.

Our classification of aquatic societies finds its basis in physiographic conditions. We recognize two principal ecological categories of aquatic organisms:

I. *Limnetic Societies*, fitted for life in the open water, and able to get along in comparative independence of the shores.

II. *Littoral Societies*, of shoreward and inland distribution.

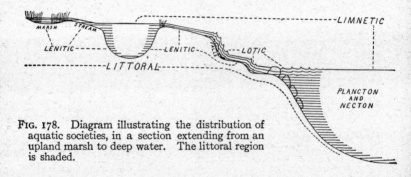

Fig. 178. Diagram illustrating the distribution of aquatic societies, in a section extending from an upland marsh to deep water. The littoral region is shaded.

The life of the open water of lakes includes very small and very large organisms, with a noteworthy scarcity of forms of intermediate size. It is rather sharply differentiated into *plancton* and *necton;* into small and large; into free-floating and free-swimming forms. These have been mentioned in Chapter V, where their main lines of adaptation were pointed out. It remains to indicate something of the composition and relations of these ecological groups.

I

PLANCTON

If one draw a net of fine silk bolting-cloth through the clear water of the open lake, where no life is visible, he will soon find that the net is straining something out

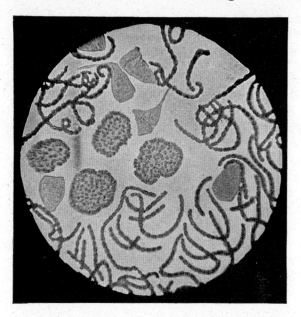

Fig. 179. "Water bloom" from the surface of Cayuga Lake. The curving filaments are algæ of the genus Anabæna. The stalked animalcules attached to the filaments are Vorticellas. The irregular bodies of small flagellate cells, massed together in soft gelatine, are Uroglenas.

of the water. If he shake down the contents and lift the net from the water he will see covering its bottom a film of stuff of a pale yellowish green or grayish or brownish color, having a more or less fishy smell, and a gelatinous consistency. If he drop a spoonful of this freshly gathered stuff into a glass of clear water and hold it toward the light, he will see it diffuse through the water, imparting a dilution of its own color; and in the midst of the flocculence, he will see numbers of minute animals swimming actively about. Little can be seen in this way, however. But if he will examine a drop of the stuff from the net bottom under the microscope, almost a new world of life will then stand revealed.

It is a world of little things; most of them too small to be seen unless magnified; most of them so transparent that they escape the unaided eye. Here are both plants and animals; producers and consumers; plants with chlorophyl, and plants that lack it; also, parasites and scavengers. And it is all adrift in the open waters of the lake.

Tho plancton-organisms are so transparent and individually so small, they sometimes accumulate in masses upon the surface of the water and thus become conspicuous as "water bloom." A number of the filamentous blue-green algæ, such as Anabæna, fig. 179, and a few flagellates, accumulate on the surface during periods of calm, hot weather. Anabæna rises in August in Cayuga Lake, and Euglena rises in June in the backwaters adjacent to the Lake (see fig. 1, on page 15).

The plants of the plancton are mainly algæ. Bacteria and parasitic fungi are ever present, though of little quantitative importance. They are, of course, important to the sanitarian. Of the higher plants there are none fitted for life in the open water; but such of their products as spores and pollen grains occur adventitiously in the plancton. It is the simply organized algæ that are best able to meet the conditions of open-water life. These constitute the producing class. These build up living substance from the raw materials offered by the inorganic world, and on these the life of all the animals of both the plancton and the necton, depends.

These are diatoms, blue-green and true-green algæ, and chlorophyl-bearing flagellates. Concerning the limnetic habits of the last named group, we have spoken briefly in Chapter IV (pp. 102–108). Being equipped with flagella, they are nearly all free-swimming. Most important among them are Ceratium, Dinobryon and Peridinium.

Most numerous in individuals of all the plancton algæ, and most constant in their occurrence throughout the year, are the diatoms (see fig. 35 on p. 111). Wherever and whenever we haul a plancton net in the open waters of river, lake or pond, we are pretty sure to get diatoms in the following forms of aggregation:

1. Flat ribbons composed of the thin cells of Diatoma, Fragillaria, and Tabelaria.

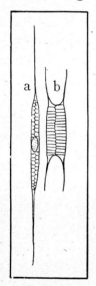

Fig. 180.
a, Rhizosolenia;
b, Attheya.

2. Cylindric filaments composed of the drum-shaped bodies of Melosira and Cyclotella.

3. Radiating colonies of Asterionella.

4. Slender single cells of Synedra.

And we may get less common forms showing such diverse structures for flotation as those of Stephanodiscus (fig. 35 *1*) and Rhizosolenia (fig. 180); or we may get such predominantly shoreward forms as Navicula and Meridion.

The blue-green algæ of the plancton are very numerous and diverse, but the more common limnetic forms are these:

1. Filamentous forms having:

(*a*)   Stiff, smoothly-contoured filaments; Oscillatoria (see fig. 34 on p. 109) and Lyngbya, etc.

(*b*)   Sinuous nodose filaments, Anabæna (fig. 179), Aphanizomenon, etc.

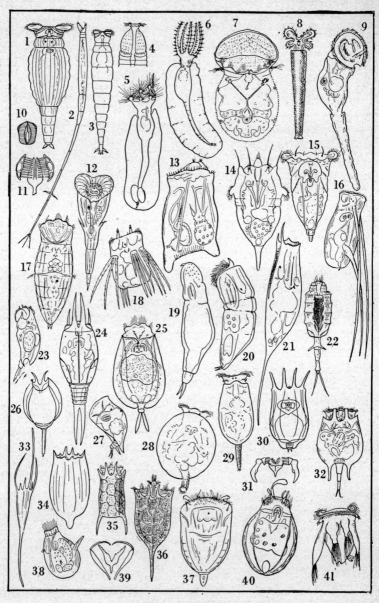

FIG. 181. Rotifers.

(*c*) Tapering filaments that are immersed in more or less spherical masses of gelatine, their points radiating outward; Gloiotrichia, Rivularia (see fig. 51, on p. 133, and 52), etc.

2. Non-filamentous forms having:

(*a*) Cells immersed in a mass of gelatine, Microcystis (including Polycystis and Clathrocystis, see fig. 51 on p. 133), Cœlosphærium, Chroococcus, etc,

b) Cells arranged in a thin flat plate. Tetrapedia (fig. 51), Merismopædia (see fig. 53 on p. 135), etc.

Representatives of all these groups, except the one last named, become at times excessively abundant in lakes and ponds, and many of them appear on the surface as "water bloom."

Of the green algæ there are a few not very common but very striking forms of rather large size found in the plancton. Such are Pediastrum (see fig. 44 on p. 123) and the desmid, Staurastum. There are many minute green algæ of the utmost diversity in form and arrangement of cells. Most of those that are shown in figure 50 on page 129 occur in the plancton; Botyrococcus is the most conspicuous of these. A few filamentous green forms such as Conferva (see fig. 45 on p. 124) and the Conjugates (fig. 41 on p. 119), occur there adventitiously, their centers of development being on shores.

The animals of the plancton are mainly protozoans, rotifers and crustaceans. The protozoans of the open

FIG. 181.

1, Philodina. 2, 3, Rotifer. 4, Adineta. 5, Floscularia. 6, Stephanoceros. 7, Apsilus.
8, Melicerta. 9, Conochilus. 10, Ramate jaws. 11, Malleo-ramate jaws. 12, Microcodon. 13, Asplanchna. 14, 15, Synchæta. 16, Triarthra. 17, Hydatina. 18, Polyarthra. 19, Diglena. 20, Diurella, 21, Rattulus. 22, Dinocharis. 23, 24, Salpina.
25, Euchlanis. 26, Monostyla. 27, Colurus. 28, 29, Pterodina. 30. Brachionus.
31, Malleate jaws. 32, Noteus. 33, 34, Notholca. 35, 36, Anuraea. 37, Plœsoma.
38, Gastropus. 39. Forcipate jaws. 40, Anapus. 42, Pedalion.

From *Genera of Plancton Organisms of the Cayuga Lake Basin*, by O. A. Johannsen and the junior author.

water are few. If we leave aside the chlorophyl-
bearing flagellates already mentioned (often considered
to be protozoa) the commoner forms among them are
such other flagellates as Mallomonas (see fig. 185 on
page 309), such sessile forms as Vorticella (fig. 179)

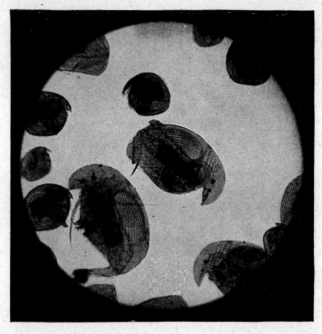

Fig. 182.　Plancton Cladocerans from Cayuga Lake.　The
larger, *Acroperus harpæ;* the smaller, *Chydorus sp.*

and such shell-bearing forms as Arcella and Difflugia
(see fig. 69 on p. 159).

　　The rotifers of the plancton are many. The most
strictly limnetic of these are little loricate forms such
as Anuraea and Notholca, two or three species of each
genus. When one looks at his catch through a microscope nothing is commoner than to see these little thin-

shelled animals tumbling indecorously about. Sometimes almost every female will be carrying a single large egg. Several larger limnetic rotifers, such as Triarthra, Polyarthra and Pedalion, bear conspicuous appendages by which they may be easily recognized. The softer-bodied Synchæta will be recognized by the pair of ear-like prominences at the front. Other common limnetic forms are shown at 2 (*Rotifer neptunius*), 21 and 25 of figure 181.

The crustacea of fresh-water plancton are its largest organisms. They are its greatest consumers of vegetable products. They are themselves its greatest contribution to the food of fishes. Most of them are herbivorous, a few eat a mixed diet of algæ and of the smaller animals. The large and powerful Leptodora is strictly carnivorous. The following are the more truly limnetic forms:

I. *Cladocerans;* species of
Daphne (fig. 234)   Diaphanosoma
Chydorus   Ceriodaphnia (fig. 165)
Bosmina (fig. 91)   Polyphemus
Sida   Bythotrephes
Acroperus (fig. 182)   Leptodora. (fig. 186)

II. *Copepods;* species of
Cyclops   Epischura
Diaptomus   Limnocalanus
Canthocamptus (see figures 95 and 96)

Of plancton animals other than those of the groups above discussed, there are no limnetic forms of any great importance. There is one crustacean of the Malacostracan group, *Mysis relicta*, that occurs in the deeper waters of the great lakes. There is one transparent water-mite, *Atax crassipes*, with unusually long and well fringed swimming legs, that may fairly be counted limnetic. There is only one limnetic insect. It is the larva of Corethra—a very transparent, free swimming larva, having within its body two pairs of air sacs that are doubtless regulators of its specific gravity.

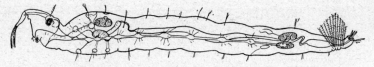

FIG. 183.   The larva of the midge, Corethra.   (After Weismann.)

Seasonal Range.   There is no period of absence of organisms from the open water, yet the amount of life produced there varies, as it does on land, with season and temperature.   In winter there are more organisms in a resting condition, and among those that continue active, there is little reproduction and much retardation of development.   Life runs more slowly in the winter. Diatoms are the most abundant of the algæ at that season.

There is least plancton in the waters toward the end of winter—February and early March in our latitude. The returning sun quickens the over-wintering forms, according to their habits, into renewed activity, and up to the optimum degree of warmth, hastens reproduction and development. With the overturn of the waters in early spring comes a great rise in the production of diatoms, these reaching their maximum oftentimes in April.   This is followed by a brisk development of diatom-eating rotifers and crustacea.   Usually the entomostraca attain their maximum for the year in May.   This rise is accompanied by a marked decline in numbers of diatoms and other algæ, due, doubtless, to consumption overtaking production.   The warmth of summer brings on the remaining algæ, first the greens and then the blue-greens, in regular seasonal succession. It brings with them a wave of the flagellate Ceratium, which, being much less eaten by animals than they, often gains a great ascendency, just as the browsing of grass in a pasture favors the growth of the weeds that are left untouched. Green algæ reach their maximum development in early summer, and blue-greens, in mid or late summer, when the weather is hottest.

With the cooling of the waters in autumn, reproduction of summer forms ceases and their numbers decline. The fall overturning and mixing of the waters usually brings on another wave of diatom production, followed by the long and gradual winter decline. This is often accompanied, as in the spring, by abundance of Dinobryon. The flagellate Synura (see fig. 30 on p. 103) is rather unusual in that its maximum development occurs often in winter under the ice.

The coming and going of the plancton organisms has been compared to the succession of flowers on a woodland slope; but the comparison is not a good one; for these wild flowers hold their places by continuously occupying them to the exclusion of newcomers. The planctonts come and go. They are rather to be likened to the succession of crops of annual weeds in a tilled field; crops that have to re-establish themselves every season. They may seed down the soil ere they quit it, but they may not re-occupy it without a struggle. And as the weeds constitute an unstable and shifting population, subject to many fluctuations, so also do the plancton organisms. They come and go; and while on their going we know that when they come again, another season, they will probably present collectively a like aspect, yet the species will be in different proportions.

There are probably many factors determining this annual distribution; but chief among them would seem to be these three:

1. Chance seeding or stocking of the waters. Each species must be in the waters, else it cannot develop there; and for every species, there are many vicissitudes (such as famine, suffocation, and parasitic diseases) determining the seeding for the next crop.

2. Temperature. Many plants and animals, as we have seen, habitually leave the open waters when they grow cooler in the autumn, and reappear in them when they are sufficiently warmed in the spring. They provide in various ways (encystment, etc.) for tiding over the intervening period. Some of them appear to be attuned to definite range of temperature. Thus the Cladoceran, Diaphanosoma, as reported by Birge for Lake Mendota, has its active period when the temperature is about 20° C. (68° F.). For this and for many other entomostraca reproduction is checked in autumn by falling temperature while food is yet abundant.

3. Available Food. Given proper physical conditions, the next requisite for livelihood is proper food. For the welfare of animal planctonts it is not enough that algæ be present in the water; they must be edible algæ. The water has its weed species, as well as its good herbs. Gloiotrichia would appear to be a weed, for Birge reports that no crustacean regularly eats it, and it is probably too large for any of the smaller animals. Birge says also ('96 p. 353), "I have seen Daphnias persistently rejecting Clathrocystis, while greedily collecting and devouring Aphanizomenon." Yet Strodtmann ('98) reports *Chydorus sphæricus* as feeding extensively on Clathrocystis, even to such extent that

its abundance in the plancton is directly related to the abundance of that alga. Each animal may have its food preference. The filaments of Lyngbya are too large for the small and immature crustaceans to handle. Ceratium has too hard a shell; it appears to be eaten only by the rather omnivorous adult Cyclops. For animal planctonts in general Anabæna and its allies and the diatoms and small flagellates appear to be the favorite food.

Obviously, the amount of food available to any species is in part determined by the numbers of other species present and eating the same things.

Plancton pulses—The organisms of the plancton come in waves of development. Now one and now another appears to be the dominant species. In most groups there are a number of forms that are competitors for place and food. The diatoms Asterionella, Fragillaria and Tabelaria may fill the upper waters of a lake together or in succession. A species of Diaptomus may dominate the waters this May, and species of Cyclops may appear in its stead next May. Yet while species fluctuate, the representation of the groups to which they belong remains fairly constant.

These sudden waves of plancton production are made possible, as every one knows, by the brief life cycle of the planctonts, and by their rapid rate of increase. If a flagellate cell, for example, divide no oftener than every three days, one cell may have more than a thousand descendants, within a month. The rotifer, Hydatina is said to have a length of life of some thirteen days, but during most of this time it is rapidly producing eggs, and the female is mature and ready to begin egg laying in 69 hours from hatching. Some of the larger animals live much longer and grow more slowly, but even such large forms as Daphne have an extraordinary rate of increase, as we have already indicated on pages 186 and 187. The rises in production grow out of:

1. Proper conditions of temperature, light, etc.
2. Abundant food
3. Rapid increase

Declines follow upon failure of any of these, and upon the attack of enemies. So swift are the changes during the growing season that those who systematically engage in the study of a lake's population take plancton samples at intervals of not more than fourteen days, and preferably, at intervals of seven days.

Local Abundance—Plancton organisms tend to be uniformly distributed in a horizontal direction. Although many of them can swim, their swimming, as we have noted in the preceding chapter, is directed far more toward maintenance of level, than toward change of location. There are, however, for many plancton organisms, well authenticated cases of irregular horizontal distribution, one of which, for Carteria, we quoted on pages 103 and 104. Alongside that record for a little flagellate, let us place Birge's ('96) record for the water-flea, *Daphne pulicaria*, in Mendota Lake.

"The Daphnias occurred in patches of irregular extent and shape, perhaps 10 by 50 meters, and these patches extended in a long belt parallel to the shore. The surface waters were crowded by the Daphnias, and great numbers of perch were feeding on them. The swarm was watched for more than an hour. The water could be seen disturbed by the perch along the shore as far as the eye could reach. * * * * On this occasion the number was shown to be 1,170,000 per cubic meter of water in the densest part of the swarm."

Shoreward Range—Few plancton organisms are strictly limited to life in open water. Most of them occur also among the shore vegetation in ponds and bays and shoals. They are very small and swim but feebly, and there is room enough for their activities in any pool. They mostly belong in the warm upper strata of the lake, and similar conditions of environment prevail in any pond. It is the deep waters of the lake that maintain uniform conditions of low and stable temperature, and scanty light; and it is the organisms of the deeper strata that do not appear in the shoals.

Hence, though the aquatic seed-plants pushing out on a lake shore are stopped suddenly at given depth, as with an iron barrier, the more simple and primitive algæ of the plancton range freely into all sorts of suitable shoreward haunts. We shall meet with them there, commingled with numberless other forms that have not mastered the conditions of the open water. In each kind of situation (pond, river or marsh has each its plancton) we shall find a different assemblage of species. In all of them we shall find the planctonts are less transparent; in none of them will there be quite such uniformity, from place to place, as is found in the population of the open waters of the lake.

Distribution in Depth. Since plancton organisms tend to be uniformly distributed in a horizontal plane one may ply his nets at any point on a lake with the expectation of obtaining a fair sample; but not so with depth, except at times when the waters are in complete circulation. A net drawn at the surface would make a very different catch from one drawn at a depth of fifty feet. Certain species found in abundance in the one would not be represented in the other. The organisms of the lakes tend to be horizontally stratified.

Each species has its own level; its own preferred habitat, where it finds optimum conditions of pressure, air, temperature and light. Fig. 184 is a diagram of the midsummer distribution in depth of seven important synthetic planctonts of Cayuga Lake.

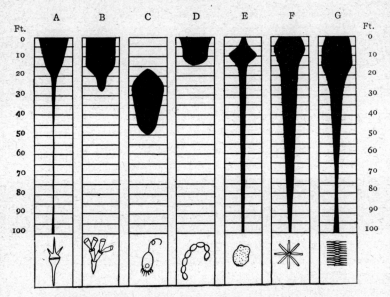

FIG. 184. Diagram illustrating midsummer distribution of seven important synthetic organisms in the first one hundred feet of depth of Cayuga Lake. *A*, Ceratium; *B*, Dinobryon; *C*, Mallomonas; *D*, Anabæna; *E*, Microcystis (Clathrocystis); *F*, Asterionella; *G*, Fragillaria.

(Based in part on Juday—15)

Light is the principal factor determining distribution in depth. This we have touched upon in Chapter II, under the subject of "Transparency." It is only in the upper strata of lakes, within the reach of effective light, that green plants can grow. Animals must likewise remain where they can find their food; whence it results that the bulk of the plancton in a lake lives in its uppermost part, the thickness of this productive stratum varying directly with the transparency of the water.

It is not at the surface, however, but usually a little below it—a depth of a meter, more or less—at which the greatest mass of the plancton is found. Full sunlight is perhaps too strong; for average planctonts a dilution of it is preferred. Free-swimming planctonts such as rotifers and entomostraca move freely upward or downward with changes of intensity of light. Anyone who has seen Daphnes in a sunlit pool congregating in the shadow of a water-lily pad will understand this. These animals rise nearer to the surface when the sun goes under a cloud, and sink again when the cloud passes. The extent of their regular diurnal migrations appears to be directly related to the transparency of the water.

Fig. 185. Mallomonas ploessi.

(After Kent.)

Temperature also is an important factor determining vertical distribution. Forms requiring the higher temperatures are summer planctonts that live at or near the surface. Others that are attuned to lower temperatures may find a congenial summer home at a greater depth. Thus the flagellate Mallomonas (fig. 185) in Cayuga Lake is rarely encountered in summer in the uppermost twenty feet of water, though it is common enough at depths between 30 and 40 feet, where the temperature remains low and constant. The average range of *Daphne pulicaria* is said to be deeper than that of other Daphnias.

The gases of the water have much to do with the distribution of animal planctonts, especially below the thermocline, where the absence of oxygen from some lakes during the summer stagnation period excludes

practically all entomostraca.   Certain hardy species of
Cyclops and Chydorus appear to be least sensitive to
stagnation conditions.   The insect Corethra, (fig. 183)
is remarkable for its ability to live in the depths, where
practically no free oxygen remains.

Age appears to be another factor in vertical distribution.   On the basis of his studies of the Entomostraca
of Lake Mendota, Birge ('96) has formulated for them
a general law of distribution, to the effect that (1)
broods of young appear first in the upper waters of the
lake (quite near the surface); (2) increase of population
results in extension downward, and the mass becomes
most uniformly distributed at its maximum development; (3) with decline of production there is relative
increase of numbers in the lower waters.

Perhaps this shifting downward merely corresponds
to the wane of vigor and progressive cessation of swimming activities with advancing age.

In the case of many plants spore development or
encystment may follow upon a seasonal wave of production, with a resulting change in vertical distribution.
Filamentous blue-green algæ develop spores.   The
ordinary vegetative filaments are buoyed up in part by
vacuoles within the cells, that lessen their specific
gravity; but spores lack these.   Hence the spore-bear-
ing filaments settle slowly to the bottom, and may be
found in numbers in the lower waters ere they have
reached their winter resting place.   Dinobryon maintains itself at the surface in part by means of the lash-
ings of its flagella, but when its cells encyst, the flagella
stop, and the fragmenting colonies slowly settle.   Thus,
both internal and external conditions have much to do
with vertical distribution.   In general it may be said
that during their period of highest vegetative activity
all plants are necessarily confined to surface waters;
that most animals are closely associated with them,

but that the constant fall of organic material toward the bottom makes it possible for some animals to dwell in the depths, if they can endure the low temperature and the other conditions found there. There are some animal planctonts, such as species of Cyclops and Diaptomus, that range the water (oxygen being present) from top to bottom. There are many that are confined during periods of activity to the warmer region above the thermocline. There are a few like Leptodora that seem to prefer intermediate depths, and there are a few (Heterocope, Limno-calanus, Mysis, etc.) that dwell in the cold water below the thermocline.

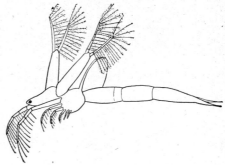

FIG. 186. Leptodora.

Collectively, this extraordinary assemblage of organisms that we know as plancton recalls in miniature the life of the fields. It has, in its teeming ranks of minute chlorophyl-bearing flagellates, diatoms and other algæ, a quick-growing, ever-present food supply that, like the grasses and low herbage on the hills, is the mainstay and dependence of its animal population. It has in some of its larger algæ the counterparts of the trees that support more special foragers, are less completely devoured, and that, through death and decomposition, return directly to the water a much larger proportion of their substance. It has in its smaller herbivorous rotifers and entomostraca, the counterpart of the hordes of rodents that infest the fields. It has in its large, plant-eating Cladocerans, such as Daphne, the equivalent of the herds of hoofed animals of the plains; and it has at least one great carnivore, that, like the tiger, ranges the fields, selecting only the larger beasts for slaughter. This is Leptodora (fig. 186). It is of phantom-like transparency, and though large enough to be conspicuous, only the pigment in its eye and the color of the food it has devoured are readily seen. It ranges the water with slow flappings of its great, wing-like antennæ. It can overtake and overpower such forms as Cyclops and Daphne and it eats them by squeezing out and sucking out the soft parts of the body, rejecting the hard shell. Leptodora, in a small way, functions in this society as do the fishes of the necton.

The total population of plancton in any lake is very considerable. Kofoid ('03) reported the maximum plancton production found by himself in Flag Lake near Havana, Ill., as 667 cubic centimeters per square meter of surface: found by Ward ('95) in Lake Michigan, 176 do.; found by Juday ('97) in the shoal water of Turkey Lake in Indiana, 1439 do. Kofoid estimated the total run-off of plancton from the Illinois River as above 67,000 cubic meters per year—this the production of the river, over and above what is consumed by the organisms dwelling in it.

If we imagine the organisms of a lake to be projected downward in a layer on the bottom, this thick layer would probably represent a quantity of life equal to that produced by an average equal area of dry land.

There is hardly another ecological group of organisms that lends itself so readily to quantitative studies, since the entire fauna and flora of the plancton may be gathered by merely straining or filtering the water. All over the world, therefore, quantitative studies have been made in every sort of lake and in many sorts of streams. Extensive data have been gathered concerning the distribution and numerical abundance of the

planctonts; but we still are sadly lacking in knowledge
of the conditions that make for their abundance.

II

THE NECTON

The large free swimming animals of the fresh waters
are all fishes. Indeed, as we have already noted (p.
233), but a few of the fishes range through the open
waters. Such are the white-fish, the ale-wife and the
ciscos,—all plancton feeders,—and a few more piratical
species, like the lake trout and the muskellonges that
feed mainly on smaller fishes.

Necton, it will thus be seen, is not a natural society.
It contains no producing class. It is sustained by the
plancton and by the products of the shores.

These fishes all have a splendid development of
stream-line form. They all swim superbly. And
according as they feed on plancton or on other fishes
they are equipped with plancton strainers or with
raptorial teeth. Excellent plancton strainers are those
of the lake fishes. They are composed of the close-set
gill-rakers on the front of the gill arches, and they
strain the water passing through. This mesh is adapted for straining the larger animal planctonts while letting the lesser chlorophyl-bearing forms slip thru.
Thus the fishes reap the crop of animals that is ma-
tured, without destroying the sources for a crop to come.

LITTORAL SOCIETIES

UNDER the sheltering influence of shores the vascular plants may grow. Animals elude the eyes of their enemies, not by becoming transparent, but by taking on colors and forms in resemblance to their environment. They escape capture, not alone by fleetness, but also by development of defensive armor, by shelter-building and by burrowing.

Large and small and all intermediate sizes occur together along shore, and those that appear betimes in open water make shifts innumerable for place and food and shelter for their young.

There are many factors affecting the grouping of littoral organisms into natural associations, most of them as yet but little studied; but the most important single factor is doubtless the water itself. The density of this medium and the consequent momentum of its masses when in motion so profoundly affect the form and habits of organisms that they may be roughly divided into two primary groups for which are suggested the following names:

I.   *Lenitic** or still-water societies.

II.   *Lotic*† or rapid-water societies, living in waves or currents.

LENITIC SOCIETIES

YOKED together, less by any common character of their own than by the lack of lotic characteristics, we include under this group name those associations of littoral organisms that dwell in the more quiet places and show no special adaptations for withstanding the wash of waves or currents. Wherever we draw the line between lenitic and lotic regions, there will be organisms to transgress it, for hydrographic conditions intergrade. We have already seen how many organisms transgress the boundary between limnetic and littoral regions. Just as in that case we found a fairly satisfactory boundary where the increasing depth of still water is such as to preclude the growth of the higher plants, so here the boundary between lenitic and lotic regions may be placed where the movement of the water is sufficient to preclude the growth of these same plants.

Lenis = calm, placid.
†*Lotus* = washed.

The reason why lenitic societies include practically the entire population of vascular plants has already been stated (p. 145): the plants have a complexity of organization that cannot withstand the stress of rapidly moving waters. They fringe all shoals, however, and they fill the more sheltered places with growths of extraordinary density. In such places they profoundly affect the conditions of life for other organisms: the supplies of food and light and air, and the opportunities for shelter.

Streams and still waters, inhabited by lenitic societies, may be divided roughly into three categories:

1. Those that are permanent.

2. Those that dry up occasionally.

3. Those that are only occasionally supplied with water.

These so completely intergrade, and so vary with years of abundance or scarcity of rainfall, that there is no good means of distinguishing between them. Perhaps for the humid Eastern States and for bodies of still water the words pond and pool and puddle convey a sense of their relative permanence. The population of the pond is, like that of the lake, to a large extent perennially active. It will be discussed in succeeding pages. That of the pool is composed of those forms that are adjusted to drouth: forms that can forefend themselves against the withdrawal of the water by migration, by encystment, by dessication, or by burrowing, or by sending roots down into the moisture of the bed. Some of these will be mentioned in the discussion of the population of the marshes. The puddles have a scanty population of forms that multiply rapidly and have a brief life cycle. The synthetic forms among them are mainly small flagellates and protococ-

Fig. 187. Lick Brook near Ithaca in spring. Its bed runs dry later in the season. (Photo by R. Matheson.)

coid green algæ.   The herbivores are such short-lived
crustaceans as Chirocephalus (see fig. 90 on p. 184) and
Apus, which have long-keeping, drouth-resisting eggs;
such rotifers as Philodina, remarkable for its capacity for
resumption of activity after dessication; such insects as
mosquitoes.   The carnivores are such adult water-bugs
and·beetles as may chance to fly into them.

Whether a population shall be able to maintain itself
depends on the continuance of favorable conditions, at
least through the period of activity of its members.
In these pages we shall give attention only to the life
of relatively permanent waters.

Plants—The shoreward distribution of plants in
natural associations is determined mainly by two
hydrographic factors: (1) movement and (2) depth of
the water.   It is directly related to exposure to waves
and to currents.   Everyone knows the difference in
appearance between plants growing immersed in a quiet

Fig. 188.   The forefront of the Canoga marshes, where partly sheltered
from the waves of Cayuga Lake, clumps of the lake bulrush lead the
advance of the shore vegetation.

pool and those growing on a wave-washed shore. The former appear as if robed in filmy mantles of green, full-fledged with leaves, and luxuriant. The latter appear as if stripped for action, unbranched, slender and bare. At one extreme are the finely-branched free-floating bladderworts (see fig. 173 on p. 285) at the other are such firmly rooted, slender, naked, pliant-topped forms as the lake bulrush (figure 188) and eel-grass. These latter anchor their bodies firmly and closely to the soil, and send up into the moving waters overhead only soft and pliant vegetative parts, that offer the least possible resistance to the movement of the water, and that, if broken, are easily replaced. The long cylindric shoots of the bulrush have their vessels lodged in the axis and surrounded with a remarkable padding of air cushions. They are not easily injured. The flat ribbon-like leaves of eel-grass are marvels of adjustability to waves.

Between these two extremes are all gradations of form and of fitness. Of the pool-inhabiting type are the water crow-foot, the water milfoil, the water horn-wort; of the opposite type are the long-leaved pond-weeds and the pipeworts. Intermediate are the broader-leaved pondweeds and Philotria.

These sometimes are found in running streams, but they usually grow in the beds in dense mutually supporting masses that deflect the current. If one place a current meter among their tops he will find little movement of the water there.

There is another place of security from waves, for such plants as can endure the conditions there. It is on the lake's bed, below the level of surface disturbance. The stoneworts (see fig. 55 on p. 137) are branched and brittle forms, very ill adapted to wave exposure, and most of them live in pools, but a few have found this place of security beneath the waves. There are extensive beds of Chara on the bottom of our great

Fig. 189. Shore-line vegetation.

lakes, at a depth of 25 feet more or less, and within the range of effective light. Associated with these, but usually on the shoreward side, are beds of pondweeds. Often there are bare wave-swept shores behind these beds with no sign of aquatic vegetation that one can see from the shore.

Depth of water determines the adjustment of aquatic seed plants in three principal categories:

1. Emergent aquatics. These occupy the shallow water, standing erect in it with their tops in the air, and are most like land plants. They are by far the most numerous in species.

2. Surface aquatics. These grow in deeper water, at the front of (and oftentimes commingling with) the preceding. The larger ones, such as the water lilies are rooted in the mud of the bottom, and bear great leaves that float upon the surface. The smaller ones such as the duckweeds (see figs. 61 and 62, p. 149) are free-floating.

3. Submerged aquatics. These form the outermost belt or zone of herbage. They are most truly aquatic in habits. Except for such forms as dwell in quiet waters, they are rooted to the bottom. Depth varies considerably within this zone. It extends from the outer limits of the preceding (hardly more than five

FIG. 189.

A. Branches of four submerged water plants: (1) Philotria, (2) Ceratophyllum, (3) Ranunculus, (4) Nais.

B. Emergent aquatics, including a clump of arrow arum; two of the pendulous club-shaped fruit-clusters are seen at (5) dipping into the water.

C. Zonal arrangement of the plants of the shore-line. The background zone is cat-tail flag (Typha). Next comes a zone of pickerel-weed (Pontederia) in full flower. Next, a zone of water lilies and such other aquatics with floating leaves as are shown in *D* and *E*. In the foreground is a zone of submerged plants—a mixture of such forms as are shown in *A* above.

D. A closer view: Lemna, free-floating and Marsilia with four parted floating leaves, and Ranunculus, in tufted sprays, submerged.

E. The floating leaves and emergent flower spikes of a pondweed, Potamogeton. (Photo by L. S. Hawkins.)

feet at most) to the limits of effective light. Within
such a range of depth conditions of movement, pressure,
warmth and light find also a considerable range; hence,
the forms differ at the inner and outer margins of the
zone. Its forefront is usually formed by Chara as
stated above, and pondweeds follow Chara, with a
number of other forms usually commingled, in the
shallower part.

These groups are not free from intergradation since
some forms like the spatterdock (fig. 195 on p. 335) are
in part emergent, and some of the pondweeds have a
few floating leaves. But they are nevertheless convenient, and they represent real ecological differences.

Distribution of these plants in depth results in their
zonal arrangement about the shore line. When all
are present they are arranged in the order indicated.
It is an inviolable order; for the emergent forms cut off
the light from those that cannot rise above the surface,
and the latter overshadow those that are submerged.
The zones may vary in width and in their component
species, but when all are present and crowded for room
they can occur only in this order. The two accompanying figures illustrate zonal arrangement; figure 189C, on
a low and marshy shore; figure 190, on a more elevated
shore, backed by a terrestrial flora.

The algæ of littoral societies are those of the plancton
(practically all of which drift into the shoals) plus
numberless additional non-limnetic forms, many of
which are sessile. As with the vascular plants, algæ
that are fragile (see fig. 198 on p. 338) and the larger
that float free (Spirogyra, etc.) develop mainly in pools
and quiet waters, while those having great pliancy of
body (Cladophora, see fig. 46 on p. 125) and protective
covering (slime-coat diatoms, etc.) are more exposed to
moving waters.

FIG. 190. Zonal arrangement of plants. At the front is a zone of Marsilea extending down into the water. Next is a zone of bur-reed, with the spiny seed-heads showing near the center of the picture. Back of this is a zone of tall composites. The flower clusters of the joe-pye-weed show above the bur-reed tops. In the background is a zone of trees.

The animal population of the shores is likewise distributed largely in relation to water movement, or to conditions resulting therefrom. There is a zonal arrangement of animal life along shores that is only a little less definite than that of plants. It is much less obvious, for plants are fixed in position and come out more into the open and into view. Nevertheless, even the most free-roving animals, the fishes, as we have already seen (p. 233), keep in the main to certain shoreward limits.

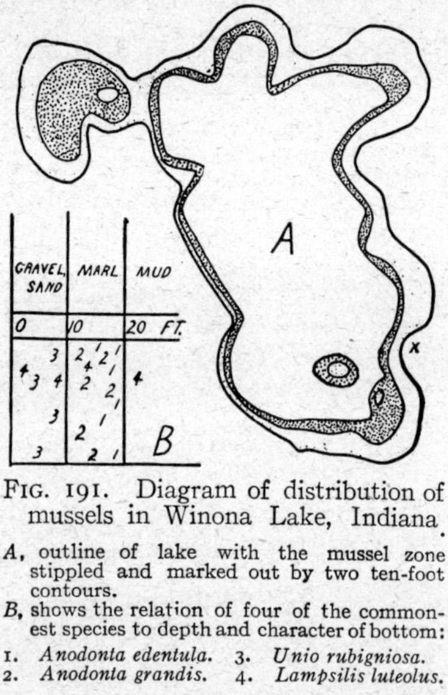

Distribution in relation to depth and to character of bottom comes out clearly in Headlee's studies of the mussels of Winona Lake. In that lake the play of the waves on shore yields a clean beach line of sand and gravel, and sifts the finer materials into deeper water. The succession is gravel and sand, marly sand, sandy marl, coarse white marl, marly mud and very soft black mud. The last named, beginning at a depth of some 20 feet, covers a very large central portion of the lake bottom. Mussels cannot live in it for they sink too deeply and the fine sediment clogs their gills. Hence the mussels are restricted to the strip along shore. Within this strip they are arranged according to hardness of bottom and exposure to waves. The accompanying diagram illustrates the distribution of four of the common species. The two Anodontas, having

FIG. 191. Diagram of distribution of mussels in Winona Lake, Indiana.

A, outline of lake with the mussel zone stippled and marked out by two ten-foot contours.

B, shows the relation of four of the common-est species to depth and character of bottom:

1. *Anodonta edentula.*   3. *Unio rubigniosa.*
2. *Anodonta grandis.*   4. *Lampsilis luteolus.*

lighter shells less prone to sink, live in the deeper zone of mixed marl and mud, and so are able to forage farther out on the bottom. On account of their thinner shells they are excluded from residence near the shore line, where the waves would crush them. The heavier shelled Unio requires a more solid bottom for its support, and is uninjured by the beating of heavy waves. Hence, its shoreward distribution. Lampsilis, however, is a more freely ranging form, having a rather light shell. It overspreads the range of all the others, coming in the less exposed places rather close to shore.

Plancton animals—The animals of the shoreward plancton are less transparent than those of the lake. They are also far more numerous. They show more color. The color is often related to situation. In small ponds and marshes they are darker as a rule than in large ponds. They include forms of very diverse habits among which are the following:

1. Forms that swim freely and continuously in the more open places. These only are common to both littoral and limnetic regions.

2. Forms that are free swimming, but that rest betimes on plants; Cladocerans with adherent "neck organs"; Copepods with hooked antennæ, etc.

3. Forms that can and that do swim betimes, but that more habitually creep on plants; many ostracods, copepods and rotifers.

4. Forms that live on or burrow in the slime that covers stems or other solid supports, and that swim but poorly and but rarely in the open water; Leeches and oligochete worms, rhizopods and midge larvæ.

5. Sessile forms that cannot swim, but that become detached and drift about passively in the open water, at certain seasons; hydras, statoblasts of fresh-water sponges and of bryozoans, resting eggs of rotifers and of cladocerans, etc.

Few of these can thrive in the waters of the limnetic region of a lake; but there is at least one member of the first group that takes advantage of an abundant supply of food in lake waters, migrates out, and develops enormously, overshadowing in numbers sometimes the truly limnetic forms.   It is *Chydorus sphæricus*.   It is rather a littoral than a limnetic species, yet it often abounds in the open lakes, following a rich development there of blue-green algæ suitable for its food.

SPATIAL RELATIONS

A large part of the animal life of the littoral region is disposed in relation to upper and lower surfaces of the water.   This grouping by levels is due to gravity. Where the air rests upon the water, making available an unlimited supply of oxygen, there at the surface are aggregated forms that require free air for breathing. Where the water rests upon the solid earth, there at the bottom are the forms that hide or burrow in the ground.

Plants and animals differ most markedly here.   Light is the prime requisite and source of energy for chlorophyl-bearing plants.   It is not light but oxygen that holds many animals at the surface of the water; and it is indifference to light that allows many other animals to dwell in the obscurity of the bottom.

Life on the bottom has a number of advantages among which are the following:

1.   Shelter is available.
2.   Energy is saved when a resting place is found, and continuous swimming is unnecessary.
3.   Gravity brings food down from above.
4.   Hiding from enemies is easier in absence of strong light.

It has also its perils chief among which are:
1.   Failure of oxygen } either of which may result
2.   Excess of silt   }   in suffocation.

In the last chapter we have discussed the more important lines of specialization that have fitted the members of the bottom population to meet or to profit by these conditions. Under the subject "pond societies," further specific illustrations will be cited.

Life at the surface is less tranquil than on the bottom. There are two kinds of animals that can maintain themselves there. (1) Those having bodies (together with the air they hold about them) lighter than the water; which rise to the surface like a cork and have to swim in order to go down below. These are mainly adult insects whose problem of getting air we have discussed in the preceding chapter.

(2) Those having bodies heavier than the water, which maintain themselves at the surface by some sort of hold on the surface film. If free-swimming, they have to swim up to the surface and break through the film before they can use it for support. Certain insect larvæ, water-fleas, rotifers, ciliates, etc., are of this habit. Creeping forms must first climb up some emergent stem, break through and then glide away suspended underneath the film. Pond-snails and hydras are of this sort. In an aquarium one may see either, hanging suspended, and dimpling the surface where the foot is attached by the downward pull on the film.

The relations of certain water-fleas to the surface film are particularly interesting. For many of these, such for example as Bosmina, this is a constant source of peril. If in swimming a Bosmina accidentally breaks through this film it falls over on its side and is held there helpless lying on the surface unable to swim away. Unless some disturbance dash it again beneath the water, its only chance for release seems to lie in moulting its skin and slipping out of it into the water. Usually when a catch of surface plancton from

Cayuga is placed in a beaker, the Bosminas begin to break through one by one, and soon are gathered in a little floating company in the center.

Scapholeberis (fig. 192), however, appears to be especially fitted to take advantage of the surface film. It is able to maintain a proper position at the surface: it possesses specialized bristles for breaking the film and laying hold upon it; its ventral (uppermost) margin is straightened and extended posteriorly in a long spine; as much contact may be had as is needed. Suspended beneath the surface, where algæ from below and pollen from the air accumulate, Scapholeberis

FIG. 192.   *Scapholeberis mucronata,*
suspended beneath the surface film.
(After Scourfield.)

rows placidly about, foraging; or it is borne along by the towing of air currents acting on the surface water—a sort of submarine sailing.

Scapholeberis is unique among water-fleas in this habit. There is also an Ostracod, Notodromas, of similar habit; and it is worthy of note that both these creatures have blackish markings on the ventral edges of the valves and are pale dorsally. As in the sloths which climb inverted in trees, the usual coloration of the body is reversed with reversal of position.

Then there are some little creatures that take advantage of the tenacity of the surface film to cover themselves with it as with a veil. Copepods, ostracods, rotifers and what not, climb up the surface of emergent stems, pushing a film of water ahead until they are well above the general surface level, where they rest and feed, and find more oxygen. The larva of Dixa is one of the most interesting of these. It will float in the surface film, but not for long, if any support be at hand. Touching a leaf it immediately bends dcuble, and pushes forward by alternate thrusts at both ends, until it has lifted a film of water to a satisfactory level.

On the surface are deposited the eggs of many insects having aquatic larvæ, but these eggs are heavier than water, and unless anchored to some solid support or buoyed up with floats (as are such eggs as those of Culex and Corethra) nearly all of them settle to the bottom. There are, however, a few midges whose egg-clusters float freely. A brief account of the egg-laying of one of them, *Chironomus meridionalis*, will illustrate several points of dependence on the surface tension.

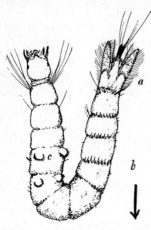

Fig. 193. Larva of a Dixa midge, inverted, to show: *a*, caudal lobe; *b*, creeping bristles; *c*, prolegs. The arrow indicates the direction of locomotion, middle foremost, both ends trailing.

The female midge, when ready to lay her eggs, rests for a time on some vertical stem by the water side in the attitude illustrated in figure 194. She extrudes her eggs which hang suspended at the tip of the abdomen. She then flies over the water carrying them securely in a rounded clump of gelatin. After a long preparatory flight, consisting of coursing back and forth in nearly horizontal lines at shoulder height above the surface of the water—a performance that lasts often twenty minutes—she settles down on the surface and rests there with outspread feet. The usefulness of her elongate tarsi is here apparent. They rest like long out-riggers radiately arranged upon the surface, easily supporting her weight while she liberates the egg mass and lets it down into the water.   At the top of the egg clump appears a circular transparent disc from which the egg mass depends. This disc catches upon the surface film, tho pulled down into it in a little rounded pit-like depression by the weight of the eggs.   Slowly the eggs descend pulling out the gelatin attaching them to the disc into a slender thread that thus becomes stretched to a length of several inches.   The female flies away to the shore and leaves them so.   Then they drift about like floating mines, transported by breezes and currents.   This little disc of gelatin dimpling the surface film is indeed a frail

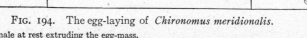

FIG. 194.   The egg-laying of *Chironomus meridionalis*.

A, The female at rest extruding the egg-mass.
B, The female resting on the surface film, letting the egg mass down into the water.
C and *D*, The egg mass liberated and hanging suspended from the surface film by a delicate gelatinous cord attached to a small disc-like float.

bark for their transportation.   When driven by waves and currents, they break their slender moorings and settle to the bottom, or adhere to floating stems against which they are tossed.

There is another phenomenon of the water surface so curious and interesting it merits passing mention here. There is a black wasp *Priocnemis flavicornis*, occasionally seen on Fall Creek at the Cornell Biological Field

Station, that combines flying with water transportation. Beavers swim with boughs for their dam, and water-striders run across the surface carrying their booty, but here is a wasp that flies above the surface towing a load too heavy to be carried. The freight is the body of a huge black spider several times as large as the body of the wasp. It is captured by the wasp in a waterside hunting expedition, paralyzed by a sting adroitly placed, and is to be used for provisioning her nest. It could scarcely be dragged across the ground, clothed as that is with the dense vegetation of the water-side; but the placid stream is an open highway. Out onto the surface the wasp drags the huge limp black carcass of the spider and, mounting into the air with her engines going and her wings steadily buzzing, she sails away across the water, trailing the spider and leaving a wake that is a miniature of that of a passing steamer. She sails a direct and unerring course to the vicinity of her burrow in the bank and brings her cargo ashore at some nearby landing. She hauls it upon the bank and then runs to her hole to see that all is ready. Then she drags the spider up the bank and into her burrow, having saved much time and energy by making use of the open waterway.

Intermediate between surface and bottom the life of the water that is not included in either of the two strata we have just been discussing, but that has continuous free range of the open water, is still considerable. It corresponds in part to the plancton of the open waters, as we have seen. It corresponds in part, also, to the necton; and, as in the open water, so also in the shoals, the larger and more important free-swimming animals are fishes. Its spatial relations are complicated by the habit some air-breathing forms (especially insects) have of ranging downward freely thro the depths, also by the way in which forms like Chironomus, that ordinarily remain in hiding in the bottom, come out betimes in the open and take a swim. But there yet remain at least two classes of organisms that belong neither to the top nor to the bottom, nor yet to the free-swimming population. These are forms that are able to sustain themselves above the mud by taking advantage of plant stems or other solid supports. These get their oxygen from the water. They are:

1. Climbing forms, that hold on by means of claws, as do the scuds and some dragonfly, damselfly and mayfly larvæ, or by a broad adhesive foot as do certain minute mussels. Many members of this group find temporary shelter between the leaves and scales of plants.

2. Sessile forms that remain more or less permanently attached, like sponges, bryozoans, hydras, etc.

Many members of both these groups construct for themselves shelters. Chironomus, for example, while usually living in such tubes as are shown in figure 134 on page 226, is able to creep about freely upon the stem. Cothurnia (fig. 73) and Stentor, and many sessile rotifers build themselves shelters.

Such support may be found on the bottom itself where that is hard; but the bottom is soft where most seed-plants grow. Furthermore, to ascend and remain above the level of the hordes of voracious bottom dwellers must be a means of safety. It is clear, therefore, that plants rising from the bottom and branching extensively must add enormously to the biological richness of the shoals, by the support and shelter they afford to such animals as these.

Size—As on land a weed patch is a miniature jungle, having a population of little insects r oughly correspond-

ing in social functions to the larger beasts of the forest,
so in the water there are large and small, assembled in
parallel associations.   The larger, as a rule, inhabit the
more open places.   Paddle-fish and sturgeons and gars
belong to the rivers; the quantative demands of their
appetites exclude them from living in the brooks   There
is not a living there for them.   Little fishes belong to
the brooks and to the shoals.   In our diagram on page
233 we have already shown how in a small lake shoreward distribution of the fishes corresponds roughly with
their size, the largest ranging farthest out, and the
smallest sticking most closely to shelter.   The senior
author has shown (07) a parallel to this in the distribution of diving beetles in an angle of the shore of a weedy
pond.   Here the most venturesome beetle was Dytiscus
(see fig. 129 on p. 221).   It was taken at the front of
the cat-tails in about three feet of water.   The associated species were disposed closely, tho not strictly in
accordance with their size, between that outer fringe
and the shore, Acilius, Coptotomus, Laccophillus,
Hydroporus, (see fig. 130) Cœlambus and Bidessus
following in succession, the last named (a mere molecule
of a beetle, having but $\frac{1}{2800}$ the weight of Dytiscus)
being found only among the trash at the very shore line.

LIFE IN SOME TYPICAL LENITIC SITUATIONS

The association of organisms in natural societies is
controlled by conditions; but conditions intergrade.
Lakes, ponds, rivers, marshes all merge insensibly,
each into any of the others; and their inhabitants
commingle on their boundaries.   Yet these names stand
for certain general average conditions that we meet
and recognize, and with which certain organisms are
regularly associated.   It will be worth while for us to
note the main characteristics of the life of several of the
more typical of such situations.

I

Pond societies—The kind of associations we now come
to discuss are typically represented in ponds, but they
occur also in any bodies of standing fresh water, that
are not too deep for growth of bottom herbage, nor too
exposed to wind and wave for the growth of emergent

FIG. 195. Where marsh and pond meet. The head of "the cove" at the Cornell Biological Field Station. Beds of spatterdock backed by acres of cattail flag. Neguena valley in the distance.

aquatics along shore. The same forms will be found
in ponds, lagoons, bayous, sheltered bays and basin-
like expansions of streams. The bordering aquatics
will tend to be arranged in zones, as discussed in the
preceding pages, according to the closeness of their
crowding.

1. The shoreward zone of emergent aquatics will
include, in our latitude, species of cat-tail (Typha), of

bur-reed (Sparganium), of bulrush (Scirpus), of spike-rush (Eleocharis), of water plantain (Alisma), of arrow-head (Sagittaria), and arrow-arum (Peltandra), of pick-erel-weed (Pontederia), of manna grass (Glyceria), etc.

2. The intermediate zone of surface aquatics will include such as:

(a). These rooted aquatics with floating leaves: white water-lily (Castalia), spatterdock (Nymphæa), water shield (Brasenia), pondweed (Potomogeton), etc.

Fig. 196. A spray of the sago pondweed, *Potamogeton*, coated with slime-coat diatoms, its leaf tips bearing dwelling tubes of midge larvæ (Chironomus).

(b). These free-floating aquatics; species of duck-weed (Lemna, Spirodela), water fernworts (Azolla, Salvinia), liverworts (Riccia), etc.

3. An outer zone of submerged plants will include such forms as pondweeds (Potamogeton), hornwort (Ceratophyllum), crow-foot (Ranunculus), naiad (Najas), eel-grass (Zostera), stonewort (Chara), etc.

These grow lustily and produce great quantities of aquatic stuff which serves in part while living, but probably in a larger part when dead, for food of the animal population, and the ultimate residue of which slowly fills up the pond. These plants contribute largely to the richness and variety of the life in the pond, by offering solid support to hosts of sessile organisms, both plants and animals. Their stems are generally quite encased with sessile and slime-coat algæ, rotifers, bryozoans, sponges, egg masses of snails and insects and dwelling tubes of midges (fig. 196). Especially do floating leaves seem to attract a great many insects to lay their eggs on the under surface. This is doubtless a shaded and cleanly place, so elevated as to be favorable for the distribution of the young on hatching.

The algæ of ponds are various beyond all enumerating. It is they, rather than the more conspicuous seed-plants, that furnish the basic supply of fresh food for the animal population. Small as they are individually, their rapid rate of increase permits mass accumulation which often become evident enough. Such are:

(1). The masses of filamentous algæ, (Spirogyra and its allies; Ulothrix, Conferva, etc.) collectively called "blanket algæ" that lie half-floating in the water, or are buoyed to the surface by accumulated oxygen bubbles.

(2). The beautiful fringes of branching sessile algæ (Chætophora, fig. 198, Cladophora, etc.) that envelop every submerged stem as with a drapery of green.

(3). The lumps of brownish gelatin inclosing compound colonies (Rivularia, see fig. 52 on p. 134, etc.), that are likely to cover the same stems later in the season, and that sometimes seem to smother the green vegetation.

(4). The spherical lumps of greenish gelatin that lie sprinkled about over the bottom—rather hard lumps inclosing compact masses of filaments of Nostoc, etc.

(5). The accumulated free-swimming forms that are not seen as discrete masses, but that tint the water. Volvox tints it a bright green; Dinobryon, yellowish; Trachelomonas, brownish; Ceratium, grayish, etc.

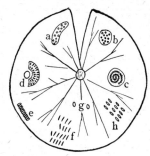

FIG. 197. Diagram of a lily-pad, inverted, showing characteristic location and arrangement of some attached egg clusters.

a, Physa; *b,* Planorbis; *c,* Triænodes; *d,* Donacia; *e,* Hydrocampa; *f,* Enallagma (inserted into punctures); *g,* Notonecta (laid singly); *h,* Gyrinus.

Such differences as these in superficial aspect, coming, as many of them do, with the regularity of the seasons, suggest to one who has studied them the principal component of the masses; but one must see them with the microscope for certain determination.

The animals of the pond that breathe free air are a few amphibians (frogs and salamanders), a few snails (pulmonates) and many insects. The insects fall into four categories according to their more habitual positions while taking air:

(1). Those that run or jump upon the surface. Here belong the water-striders and their allies—long legged insects equipped with fringed and water-repellent feet that take hold on the surface film, but do not break through it. Here belong many little Diptera that rest down upon the surface between periods of flying.

Here belong the hosts of minute spring-tails that gather in the edges in sheltered places, often in such numbers as to blacken the surface as with deposits of soot. Minute as these are they are readily recognized by their habits of making relatively enormous leaps from place to place.

(2). Those that lie prone upon the surface. Best known of these because everywhere conspicuous on still

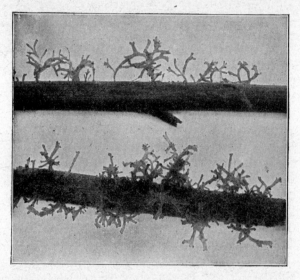

FIG. 198. Two fallen stems enveloped with a rich growth of the alga, *Chætophora incrassata.*

waters, are the whirl-i-gig beetles. Less common and much less conspicuous are the pupæ of the soldier-flies (Stratiomyia, etc.) and the larvæ of the Dixa midges.

(3). Those that hang as if suspended at the surface, with only that part of the body that has to do with intake of air breaking through the surface film. Here belong by far the larger number of aquatic insects. Here are the bugs and the adult beetles, alertly poised, with oar-like hind legs swung forward, ready, so that a stroke will carry them down below in case of approach of danger. Here hang the wrigglers—larvæ and pupæ of mosquitoes. Here belong the more passive larvæ of many beetles and flies and the pupæ of swale-flies and certain crane-flies.

(4). Those that rest down below, equipped with a long respiratory tube for reaching up to the surface for

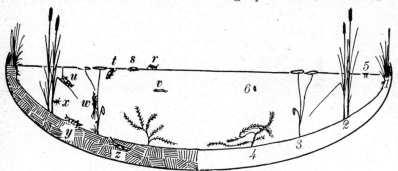

FIG. 199. Diagram of distribution of pond life. The right side illustrates the zonal distribution of the higher plants. *1*, shore zone; *2*, standing emergent aquatics; *3*, aquatics with floating leaves; *4*, submerged aquatics; *5*, floating aquatics; *6*, free swimming algæ of the open water.
The left side represents the principal features of the distribution of animals. *r, s, t, u*, forms that breathe air; *v, w, x, y*, and *z*, forms that get their oxygen from the water.

(From the Senior Author's *General Biology*)

air. Such are Ranatra, and the rat-tailed maggots of syrphus-flies.

The animals of the pond that are more strictly aquatic in respiratory habits (being able to take their oxygen supply from the water itself) are so numerous that we shall be able to mention only a few of the larger and more characteristic forms. First there are the inhabitants of the bottom. These fall into two principal categories, the free-living and the shelter-building forms. The free-living forms may be grouped as follows:

(1). Bottom sprawlers that lie exposed, or only covered over with adherent silt. These are character-ized by a marked resemblance to their environment. Such crustaceans as the crawfish and Asellus, such insects as Ephemerella, Cænis and other mayfly nymphs Libellula, Didymops, Celithemis (fig. 200) and other dragonfly nymphs, and certain snails and flatworms belong here.

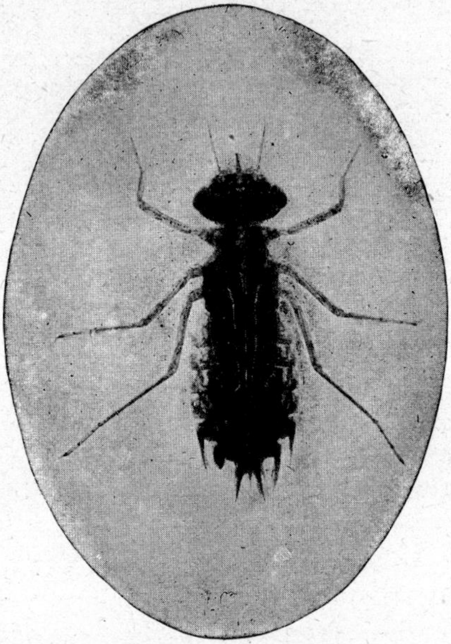

FIG. 200. A bottom sprawler: nymph of of the dragonfly, *Celithemis eponina*.

(2). Bottom dwel-lers that descend more or less deeply into the mud or sand, by the various means already discussed in the pre-ceding chapter. Among the shallow burrowers are many shell-bearing molluscs, both mussels and snails; a few may-fly and dragonfly nymphs. Descending more deeply in muddy beds are some true worms and horsefly larvæ.

The shelter-building forms of the bottom may be grouped as:

(1). Forms making portable shelters. These are mainly caddis-worms that construct cases of pieces of wood or grains of sand.

(2). Forms making fixed shelters. These are such caddis-worms as Polycentropus, such worms as Tubifex (see fig. 83 on p. 174) and such midges as Chironomus (see fig. 134 on p. 220).

It is some of these animals of the pond bottom that give to the littoral region its great extension out under the open waters of the lakes. It is only a few members of the population that are able to endure conditions in the depths far out from shores. These are such as:

Small mussels of genus Pisidium.

Mayfly nymphs of the genus Hexagenia.

Midge larvæ of the genus Chironomus.

Caddis-worms in the cylindric cases of sand, not yet certainly identified, etc.

The larger animals of the pond that belong neither to surface nor bottom and that correspond to neither plancton nor necton of the open water may be grouped as:

(1). Climbing forms (most of which can swim on occasion), such as the scuds (Amphipods), the nymphs of dragonflies such as Anax, of damselflies such as Lestes and Ischnura, of mayflies such as Callibætis, larvæ of caddisflies such as Phryganea and of moths such as Paraponyx, mussels such as Calyculina, and many leeches, entomostracans and rotifers.

(2). Sessile forms such as hydras, sponges, bryzoans and rotifers.

II

Marsh Societies.—We come now to consider the associations of organisms in waters that are not too deep for the growth of standing aquatics. Shoalness of water and instability of temperature and other physical conditions at once exclude from residence in the marsh the plants and animals of more strictly limnetic habits; but it is doubtless the presence of dense emergent plant growth that most affects the entire population. This gives shelter to a considerable number of the higher vertebrates, and these rather than the fishes are the large consumers of marsh products. The muskrat

breeds here and builds his nest of rushes.   He prefers,
to be sure, the edge of a marsh opening, where in deep
water he may find crawfishes and molluscs, with which
to vary his ordinary diet of succulent shoots and tubers.

FIG. 201.   The eggs of the spotted salamander, Ambystoma punctatum.
(Photo by A. A. Allen.)

Deep in the marsh dwell water birds, such as grebes,
rails, coots, terns, bitterns, and in the north, ducks and
geese as well.   Such non-aquatic birds as the long-billed
marsh-wren and the red-winged blackbird use the top
of the marsh cover as a place to build their nests and

use also the leaves of marsh plants for building materials. Several turtles and water snakes are permanent residents as are also a few of the frogs. Most of the frogs visit the marsh pools at spawning time, making the air resound with their nuptial melodies. The spotted salamander is the earliest amphibian to spawn there. Though the adult is but a transient, its larvæ remain in the marsh pools through the season.

The plants are the same kinds found in the marginal zone of the pond border, but here they often cover large areas in a nearly pure stand. In

FIG. 202. Tear-thumb.

our latitude in the more permanent waters, the dominant species usually are cat-tail, phragmites, bur-reed and the soft-stemmed bulrushes; in the shoals that dry up each year they are sweet flag, sedges, manna grass and the hard-stemmed bulrushes. Such plants as these have strong interlaced roots and runners that form the basis of the marsh

cover, and that support a considerable variety of more
scattering species.   One of the most widespread of
these secondary forms is the beautiful marsh fern,
whose black rootstocks over-run the tussocks of the
sedges, shooting up numberless fronds.   Scattering
semi-aquatic representatives of familiar garden groups
are the marsh bellwort (*Campanula aparinoides*), the
marsh St. John's wort (*Hypericum virginicum*) and the
marsh skull-cap (*Scutellaria galericulata*): these are
dwarfish forms, however, that nestle about the bases of
the taller clumps.   With them are straggling prickly
forms, such as the marsh bedstraw (*Galium palustre*),
the white grass (Leersia) and the tear-thumb (*Polygonum sagittatum*, fig. 202).   Strong growing forms that
penetrate the marsh cover with stout almost vine-like
stems are the marsh five-finger (*Potentilla palustris*) the

FIG. 203. A marshy pool with flowers of the white water crow-foot rising
from the surface.

joint weed (*Polygonum*) and the buck-bean (*Menyan-thes trifoliata*). True climbers also, are present in the marsh although usually only on its borders; such are the climbing nightshade bittersweet (*Solanum dulca-mara*) and the beautiful fragrant-flowered climbing hemp-weed (*Mikania scandens*). Here and there one may see a protruding top of swamp dock (*Rumex verticillatus*), a water hemlock (*Cicuta bulbifera*) or a swamp milkweed (*Asclepias incarnata*).

Every opening in the marsh contains forms that are more characteristic of ponds and ditches, such as arrow-heads and water plantain. And even the little trash filled pools often contain their submerged aquatics. Such a one is shown in the figure 203, a shallow pool filled with fallen leaves, its surface suddenly sprinkled over with little star-like flowers when the white water-crowfoot shoots up its blossoms.

Algæ often fill these pools; sometimes minute free-swimming forms that tint their waters, but more often "blanket algæ," whose densely felted mats may smother the larger submerged aquatics.

The animal life of the marsh is also a mixture of pond forms and of forms that belong to the more permanent waters. The fishes are bullheads and top minnows and others that can endure foul waters, scanty oxygen and rapid fluctuations of temperature. Of crustaceans, ostracods and scuds are most abundant. Of molluscs, Pisidium and Planorbis are much in evidence, and other snails are common. Insects abound. Some are aqua-tic and some live on the plants. Of all Odonata, Lestes (fig. 204) is perhaps the most characteristic marsh inhabi-tant; of mayflies, Blasturus and Cænis; stoneflies, there are none. Of caddis-flies there are many, but *Limno-philus indivisus* is perhaps the most characteristic marsh species. It is not known to inhabit any waters except those that dry up in summer. The commonest beetles are small members of the families Hydrophilidæ, Dytiscidæ and Haliplidæ. The most characteristic of the bugs is the slender little marsh-treader, Limnobates. Swale-flies, mosquitoes, crane-flies and ubiquitous midges abundantly represent the aquatic Diptera.

FIG. 204. A damselfly. *Lestes uncatus.*

There are, of course, many insects dependent upon particular plants. Such are the tineid moth, *Limnacea phragmitella*, that burrows when a larva in the Typha fruit spike, and the weevil, *Sphenophorus*, that burrows in the Typha crown; the leaf-beetle, *Donacia emerginata*, whose larva feeds on the submerged roots of the bur-reed, etc. Here are also a number of characteristic spiders, such as the diving spider, Dolomedes.

Doubtless the lower groups of animals possess species that are addicted to dwelling in marshes, and fitted to the peculiar conditions such places impose, but these have been little studied. There is hardly any situation where the fauna is so imperfectly known.

As compared with the land, fauna and flora of marshes are characterized by a small number of species, and enormous numbers of individuals. In other words,

FIG. 205. "Tree-swallow pond": a once famous collecting ground in the Renwick marshes at Ithaca. Photo taken in spring after the burning and the freezing and the floods, but before the growth of the season.

the population is one of small variety but of great density. Such forms as are fitted to maintain themselves where floods and fire alternately run riot find in the rich soil and abundant light and moisture opportunity for a great development. Fire sweeps the surface clear of trees, which would overtop and overshadow the herbage and would create swamp conditions. The ground layer of water-soaked trash prevents the burning of the root stocks; it also prevents deep freezing after the fires have run. Plants that are capable of renewing their vegetative shoots from parts below the level of the burning, are the ones that year after year, maintain their place in the sun.

III

Bog Societies. Bogs belong to moist climates and to places where water is held continuously in an amount sufficient to greatly retard the complete decay of plant remains. Acids accumulate, especially, humous acids. The soil becomes poor in nutriment, especially in available nitrogen. Plants can absorb little water, at least at low temperature; and the typical bog situation is therefore said to be "physiologically dry." With such conditions there go some striking differences in flora and fauna. The plants are "oxylophytes" like sphagnum and cranberry, i. e., plants that can grow in more or less acid media, and that have many of the superficial characteristics of desert plants; such as vestiture of hairs or scales or coatings of wax, thickened cuticle, leaves so formed or so closed together as to limit or retard transpiration. The kinds of plants are fewer; the individuals crowd prodigiously. They are eaten by animals less than in any other situation. Their remains, partly decomposed, are added to the soil in the form of deposits of peat. The animal population is correspondingly reduced and scanty.

Sphagnum. The most characteristic single organism in such a situation is the bog-moss, Sphagnum (fig. 206; see also fig. 59 on p. 147). This grows in cushion-like masses of soft erect unbranched stems, that are individually too weak and flaccid to stand alone, but that collectively make up the largest part of the bog cover. The masses are loose and easily penetrated by the roots and runners of other stronger plants. It is the inter-penetration of these that binds the bog cover together, making it resilient under foot.

The leaves of Sphagnum are interspersed with cells that are mere water reservoirs having porous walls. Some of these leaves are deflexed against the stem and make excellent capillary conduits for water upward or downward. Whether the abundant supply be in the air above or in the soil below, these make provision for the equitable distribution of it. Wherefore, these masses of sphagnum become water reservoirs, holding their supply often against gravity, and bathing the roots of all the cover plants that rise above the surface of the bog.

Sphagnum belongs to the shore, and it is quite incapable of advancing into the water unassisted. But with

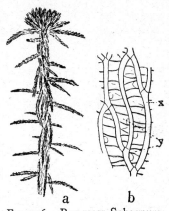

FIG. 206. Bog moss, Sphagnum; *a* the tip of a spray; *b*, a few cells from a leaf; *x*, long interlaced lines of slender sinuous chlorophyl-bearing cells, and *y*, large empty water reservoir cells having pores in their walls for admission of water and annular cuticularisations for support.

the help of stronger more straggling plants whose roots and branches penetrate and interlace in its masses in mutual support, it is able to extend as a floating border out over the surface of still water in small lakes and ponds. These floating edges may be depressed by the weight of a man until they are under water, but they are tough and elastic, and rise again unharmed when the weight is removed. Long, strong, pliant-stemmed heaths and slender sedges are the plants commonly associated with sphagnum in the making of this floating border. In the bog cover equally close is its association with the common edible cranberry.

Some habitual associates of sphagnum are shown in figure 207. In such a place as the foreground of this picture, if one slice the bog cover with a hay-knife, he

FIG. 207. A bit of bog cover. (McLean, N. Y.). From the central clump of pitcher-plant leaves rises one long-stalked flower. The surrounding bog moss is Sphagnum. A few slender stems of cranberry trail over the moss. The taller shrubs are mainly heaths such as Cassandra and Andromeda.

(Photo by H. H. Knight.)

may easily lift up the slices; for they are composed of living material to a depth of only about a foot. Below is peat; at first light colored and composed of identifiable plant remains, but, deeper, becoming darker and more completely disintegrated. The slices cut from the surface have sphagnum for their filling, but they are tough and pliant, like strips of felt, owing to the close interlacing of roots and stems of the other plants of the bog cover.

Many delightful herbs grow on the surface of the bog. The pitcher-plant shown in our figure is one, and the sundew (see fig. 172 on p. 283) is another carnivorous species. These, as we have seen in the preceding chapter, have their own way of getting nitrogen when the available supply is small. Orchids of several genera (Habenaria, etc.) and moccasin flowers (fig. 208) there bear beautiful flowers. Cotton grass (Eriophorum) is showy enough with its white tufts held aloft when in fruit, and a beaked rush (Rynchospora) is its natural associate. In places where the surface rises in little hummocks, there are apt to be patches of the xerophytic moss, Polytrichium, associated with charming little colonies of wintergreen and goldthread. At the rear of the heath shown in our figure stand huckleberries and bog brambles and masses of tall bog ferns while thickets of alder and dogwood crowd farther back.

Fig. 208.   A charming bog plant, the moccasin flower. (*Cypripedium reginae*).

Where sphagnum borders on open water, there often lies in front of it the usual zone of aquatics with floating leaves, as shown in the accompanying picture, and in

still deeper water there are apt to be beds of Chara and
of pondweeds. These and the molluscs associated
with them, leave their calcified remains deposited on
the pond bottom as a stratum of marl. Thus the

FIG. 209. Mud pond, near McLean, N. Y. This is a bog pond, surrounded
in part at least by floating sphagnum. The outlet (to left in the picture) is
bordered by tussock sedges, backed up by extensive alder thickets.

(*Photo by John T. Needham.*)

filling of a bog pond is in time accomplished by the
deposition of a layer of marl over its bottom, and a
much thicker mass of peat over the marl. Successive
stages in the filling process are graphically shown in
Dachnowski's diagram, copied on the next page.

Peat formation and filling of beds goes on, of course, in
ponds where there is no sphagnum; goes on wherever

the conditions for incomplete decay of plants prevail; and from the foregoing it will be seen that peat is not likely to be composed of the remains of sphagnum alone. The forefront of advancing shore vegetation is led by a number of plants of very different character.

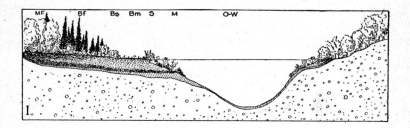

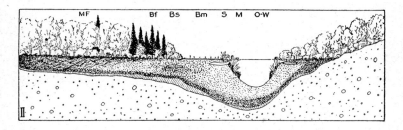

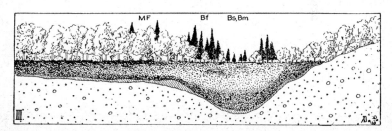

FIG. 210. Dachnowski's diagram illustrating three stages in the filling of a pond with deposits of peat and marl. Peat is stippled; marl, cross-lined.

OW, open water; M, marginal succession; S, shore succession; B, bog succession, including bog meadow (Bm), bog shrub (Bs), and bog forest (Bf); MF, mesophytic forest.

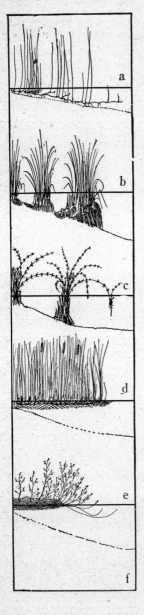

The accompanying diagram shows five modes of progress into deeper water of pioneer land-building plants.

a is the method of the spike-rush on gently sloping shore. It is the method by which numberless shore plants extend their holdings,—subterranean off-shoots.

b is the method of the tussock sedges (see also fig. 209) which on the loose mud in shallow waters build up solid clumps. Many of these, less than a foot in diameter, are yet of such firmness that they will sustain the weight of a man. Every one knows such clumps, from having used them (as stepping stones are used) in crossing a swale. New offsets lie hard against the old ones, roots descend in close contact, and fibrous rootlets interlace below in extraordinary density.

c is the method of the swamp loosestrife, Decodon, a method of advancing by long single strides. The tips of the long over-arching shoots dip into the water, and then develop roots and buds and a copious envelope of aerating tissue. If these new roots succeed

FIG. 211. Diagram illustrating the method of advance into deeper waters of typical land-building plants.

a, Spike rush; b, tussock sedge; c, swamp loosestrife; d, cat-tail flag; e, Sphagnum and heath s.

in taking a good hold on the bottom, then other shoots spring from this new center and repeat the process.

d is the method of the cat-tail flag. It consists in developing an abundance of interlaced fibrous roots, and then simply floating on them. Much mutual support is required by plants that grow so tall; and any great advance of a few clumps beyond the general front may result in disaster from overturn by winds.

e is a method of mutual support between species of very different sorts. It is that of the sphagnum and heaths just discussed. Greater progress over deep water is made by this method than by any of the others.

A photograph of the first named is reproduced as figure 212.

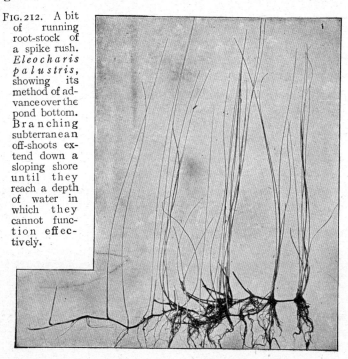

FIG. 212. A bit of running root-stock of a spike rush. *Eleocharis palustris,* showing its method of advance over the pond bottom. Branching subterranean off-shoots extend down a sloping shore until they reach a depth of water in which they cannot function effectively.

IV

The population of stream beds—If we distinguish
between lenitic and lotic societies by presence or
absence of growths of vascular plants, then the greater
part of stream beds shelter lenitic societies. The
greater part has not a current of sufficient swiftness to
prevent the growth of such plants. And indeed it is
only in restricted portions of any stream that we
find the animals specially adapted to meet conditions
imposed by currents.

Where the stream bed forms a basin, there the conditions of life, for the larger organisms at least, approximate those of a lake. Hence we find in those places
in large streams where the water is deep and still, there
occur many forms like those in lakes. The sturgeon
belongs in both, and so do the big mussels and the
operculate snails, the big burrowing mayflies, the big
tube dwelling midge larvæ, etc. The basins of creeks
offer conditions like those in ponds; the basins of
brooks, conditions like those of pools. And the largest
species are restricted to such of the larger basins as can
afford them adequate pasturage and suitable places for
rearing their young. To be sure, in all those basins,
the water is constantly passing on down stream and
the plancton of the basin, while in part developing there,
is in a large part constantly lost below and constantly
renewed from above. Kofoid (o8) states that "The
plancton of the Illinois River is the result of the
mingling of small contributions by tributary streams,
largely of littoral organisms and the quickly growing
algæ and flagellates, and of the rich and varied plancton
of tributary backwaters, present in an unusual degree
in the Illinois because of its slightly developed flood-
plain, and from which it is never entirely cut off, even
at lowest water. * * * * To these elements is added
such further development of the contributed or indigen-

ous organisms as time permits, or the special conditions of nutrition and sewage contamination facilitate. Though continually discharging, the stream maintains the continuous supply of plancton, largely by virtue of the reservoir backwaters—the great seedbeds from which the plancton-poor but well fertilized contributions of tributary streams are continuously sown with organisms whose further development produces in the Illinois River a plancton unsurpassed in abundance."

Doubtless, in every stream the plancton supply is constantly renewed from sheltered and well populated basins, which serve as propagating beds. And, indeed, on every solid support diatoms are growing, and the excess of their increase is constantly being released into the passing current. In the swiftly flowing, plancton-poor streams about Ithaca there is not time for much increase of free planctons by breeding. The waters run so swift a course they can only carry into the lake such forms as they have swept from their channels in their rapid descent.

While there has been much study of the life of the open waters of rivers there has hitherto been little study of their beds. Where the beds are sandy with flow of water over them we know the life differs from that of muddy basins. The heavier-shelled mussels and snails are on the sand; and the commoner insects there are the burrowing nymphs of mayflies and Gomphine dragon-flies, and the caddis-worms that live in portable tubes of sand.

The beds of the smallest streams are easy of access, and a few observations are available to indicate that their study will bring to light some interesting ecological relations. A few very restricted situations will be cited in illustration.

FIG. 213. A moss-bed covering the face of a rock ledge (in flood time, a waterfall) in the bed of Williams Brook at Ithaca, N. Y. The water seen on the rock above trickles down through this moss. Here is a restricted and peculiar animal population.

Moss patches—On the rocky beds of large brooks that run low but do not entirely run dry, there are frequent patches of the close-growing m o s s , Hydrohypnum. These patches frequently cover the vertical face of a waterfall (fig. 213). The little water that remains in dry season trickles through the layer of moss, and in times of flood the speedier torrent jumps over it. Under the flattened frond-like green sprays there is comparatively quiet water at all times; and in this situation there lives a peculiar assemblage of insects that differ utterly from the lotic forms dwelling in the same streams (to be discussed in a later part of this chapter), tho often dwelling within a few feet of them. They lack all the usual adaptations for meeting the wash of currents. They are (with occasional intermixture of a few larvæ of small midges and of Simulium) the following:

1. The slender larvæ of soldier-flies (*Euparhyphus brevicornis*). Each bears a pair of ventral hooks that may serve for attachment.

2. The greenish larvæ of the cranefly (*Dicranomyia simulans*).

3. The warty-backed larvæ of the Parnid beetle (*Elmis quadrinotatus*).

4. Larvæ and pupæ of a little black Anthomyid fly (*Limnophora* sp.?).

a   b   c   d   e   f   g

FIG. 214. Insect larvæ from a moss patch such as is shown in the preceding figure. *a*, Psychoda; *b*, Elmis; *c*, *d*, *e*, Euparhyphus, *c*, being lateral, *d*, dorsal and *e*, ventral views, *c* and *e* show the huge ventral hooks on the penultimate segment; *f* and *g*, cases of an unknown caddis-worm, *f*, composed mainly of sand; *g*, mainly of moss.

5.   The slender larvæ of a moth fly (*Psychoda alternata*), its body covered with deflexed spines.

6.   The larvæ of an unknown caddis-fly whose cases are composed sometimes of stones, sometimes of moss fragments.

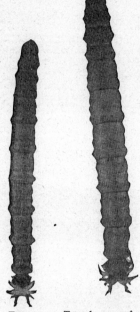

FIG. 215.   Two larvæ of the giant cranefly, *Tipula abdominalis*, an inhabitant of leaf-drifts in woodland brooks.
Natural size.

Leaf-drifts—In the beds of woodland brooks, there are barriers of fallen leaves, piled by the current upon the bare, obtruding roots of trees.   These leaf-drifts have a population of their own, the most charactertistic member of which about Ithaca is the huge larva shown in figure 215.   This is the larva of the giant cranefly, *Tipula abdominalis*. Associated with this larva in these water-soaked masses of leaves, are the nymphs of such stoneflies as Nemoura and of such mayflies as Bætis and Leptophlebia, a few beetles and often many scuds (Gammarus).   In the mud behind the leaf-drifts, there are often earthworms, washed down from fields above.

In the clear pools in upland streams that flow through swampy woods, when the bottom is strewn with forest litter intermixed with brownish silt, there dwell a number of forms that certainly belong to the lenitic rather than to the lotic societies. Such are the caddisworms of figure 216.   With these are associated small mussels of the genus Sphærium, squat dragonfly nymphs of the genus Cordulegaster, and climbing nymphs of the genus Boyeria, water-skaters on the surface and burrowing mayflies in the beds, and a considerable variety of the lesser midges on every possible support.

We have already noted (page 86) that slack water exists behind boulders and other obstructions in the bed of rapid streams: but this is not stagnant water; and the animals living in such shelter, if the current above them be swift, are hardly ever of the same species that are found in ponds. Only in slow-flowing waters, where conditions merge, do lotic and lenitic forms become near neighbors.

FIG. 216. A bit of the bed of a pool in a woodland stream showing among the forest litter the wooden cases of the larva, of the caddis-fly, *Halesus guttifer*. (See also fig. 104 on p. 198.) Protective resemblance. There are 14 cases in the picture.

A DIAGRAM OF LOCALIZATION OF INSECTS IN THE BED OF A SWIFT STREAM

(From Needham and Christenson, '27)

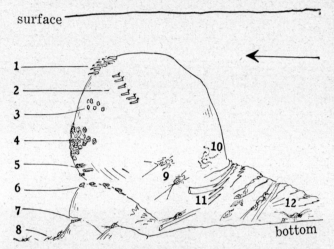

Twelve situations about a boulder, and the insects commonly found in them in Logan River, Utah, are as follows:

1. Simulium; black fly larvae; fully exposed where current is swiftest.
2. Brachycentrus; caddis worms; in square cases attached by the upstream 'end.
3. Bibiocephala; net-winged midge-larvae; of limpet-form, adhering by ventral suckers.
4. Glossosoma; caddis worms, in pebbly cases on down-stream face of boulders.
5. Antocha; carnivorous cranefly larva; in tubes on down-stream face of boulder.
6. Hydropsyche; net-spinning caddis worm, making nets beside a crevice where water breaks over.
7. Atherix; snipe fly larva, living in crevices.
8. Baetis and Leptophlebia; mayfly nymphs, living on bottom in slackened current.
9. Iron and Rithrogena; mayfly nymphs, clinging to broad surfaces, mostly underneath.
10. Chironomus and Tanytarsus; midge larvae, living in tubes in more or less exposed places.
11. *Ephemerella grandis*, the prickle-back mayfly nymph; clinging to trash in half sheltered places.
12. Acroneuria and Pteronarcys; stonefly nymphs; living amid the trash and sheltered by it.

The arrow indicates direction of the current.

LOTIC
SOCIETIES

ACCORDING to the grouping outlined on page 315, we designate by this name those assemblages of organisms that are fitted for life in rapidly moving water —that are washed by currents, as the name signifies. Whether the water flow steadily in one direction as in streams, or back and forth with frequent shifts of direction as on wave-washed shores, the organisms present in it will be much the same sorts. The plants will be mainly such algæ as Cladophora, and slime-coat diatoms: the animals will be mainly net-spinning caddis-worms and a variety of more or less limpet-shaped invertebrates.

The animals of lotic societies are mainly small invertebrates. There are fishes, indeed, like the darters that live in the beds of rapid streams. These lie on the bottom where the current slackens, lightly poised on their large pectoral fins, or rest in the lee of stones, darting from one shelter to another. It is only a few lesser animals, of highly adapted form and habits, that are able to dwell constantly in the rush of waters.

These lesser animals may be roughly divided into two categories according to the sources of their principal food supply:

1.   Plancton gathering forms, that are equipped with an apparatus for straining minute organisms out of the open current.

2.   Ordinary forms that gather home-grown food about their dwelling places.

1.   *Plancton Gatherers.*—These are they that live mainly on imported food, which by means of nets or baskets or strainers they gather out of the passing current.   These are the most typical of lotic organisms, for they must needs live on the exposed surfaces that are washed by the current.   They dwell on the bare rock ledge, over which the water glides swiftly, or on the top of the boulders in the stream bed, or on the exposed side of the wave-washed pier.   They are few in kinds, and very diverse in form, and show many signs of independent adaptation to life in such situations. Among them are four that occur abundantly in the Ithaca fauna.   These four and their mode of attachment and of plancton gathering are illustrated in the accompanying diagram.   The fly larva, Simulium, adheres by a caudal sucker, gathers plancton by means of a pair of fans placed beside its mouth, while its body dangles head downward in the stream.   The larva of the caddis-fly, Hydropsyche, lives in a tube and constructs a net of silk that strains organisms out of the water running through it.   The caddis-worm, Brachycentrus, attaches the front end of its case firmly to the top of a boulder in the stream bed, and then spreads its bristle-fringed middle and hind feet widely to gather in any organisms that may be adrift in the passing water.

The nymph of the "Howdy" Mayfly, Chirotenetes, fixes itself firmly with the stout claws of its middle and hind feet clutching a support, and extends its long fore feet with their paired fringes of long hair outspread like a basket to receive what booty the current may bring. These four are so different they are better considered a little further separately.

The larva of Simulium (the black-fly, or buffalo gnat) perhaps the most wide-spread and characteristic animal

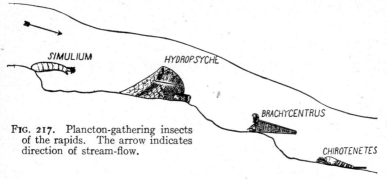

FIG. 217. Plancton-gathering insects of the rapids. The arrow indicates direction of stream-flow.

of running water, is unique in form and in habits. It hangs on by means of a powerful sucker that is located near the caudal end of its soft and pliant bag-shaped body. But it may also attach itself to the stones by a silken thread spun from its mouth: and if it then loosens its sucker, it will dangle at the end of the thread, head upstream. By means of these two attachments, it may travel from place to place without being washed away, but in the swiftest water, it can make only short moves sidewise. It travels by loopings of its body, like a leech. So it shifts its location with changes of water level, always seeking the most exposed ledges which a thin sheet of water pours over. There it gathers in companies, so closely placed side by side as to form great black patches on the stones.

There is little movement from place to place. The larvæ hang at full stretch, their pliant bodies swaying with little oscillations of the current, their fans outspread, straining what the passing stream affords. Each of these fans is composed of several dozen slender rays, each one of which is toothed along one margin like a comb of microscopic fineness, and all have a parallel curvature like the fingers of an old-fashioned reaper's cradle. They are efficient strainers.

When grown the larva spins its half cornucopia-shaped straw-yellow cocoon on the vertical face of a ledge where the water will fall across its upturned open end, then transforms to a pupa inside. The pupa bears on the prothorax a pair of long, conspicuous, many branched respiratory horns, or "tube gills" (see fig. 171 on p. 280).

The eggs are laid at the edge of the swiftly flowing water on any solid support, on the narrow strip that is kept wet, and, by oscillations of the current occasionally submerged.

Hydropsyche, the seine making caddis-worm, lives in sheltering tubes of silk, spun from its own silk glands, fixed in position on the surface of a stone (oftenest in some crevice), and covered on the outside with attached sticks or broken fragments of leaves or stones. Always one end of the tube is exposed to the current, and at this end, the larva reaches out to forage. Here it constructs its net of crosswoven threads of fine silk. The net is a more or less funnel-shaped extension of silk from the front of the dwelling-tube. The opening is directed upstream, so that the current keeps it fully distended. The semi-circular front margin is held in place by means of extra staylines of silk. The mesh is rather open on the sides, but on the bottom there is usually a small feeding surface that is much more closely woven.

The larva lies in its tube in readiness to seize anything the current may throw down upon its feeding surface or entangle in the sides of its net. The whole net is so delicate that it collapses on removal from the water. To see it in action, it is best examined through a "water-glass."*

Brachycentrus, the "Cubist" caddis-worm, is restricted in habitat to spring-fed streams flowing through upland bogs. It constructs a beautiful case that is square in cross-section. Each side is covered with a single row of sticks (bits of leaf stalks, grass stems, etc.) placed crosswise. The larva fastens its case by a stout silken attachment to the top of some current-swept boulder and then rests with legs outspread as indicated in figure 217 in a receptive attitude, waiting for whatever organic materials the current may bring within its grasp.

The Nymph of Chirotenetes, the "Howdy" Mayfly, lives on the rock ledge or where the water sweeps among the stones. Its body is of the stream-line form discussed in the last chapter—the form best adapted to diminishing resistance to the passage of water, as well when at rest as when swimming. The nymph sits firmly on its middle and hind feet. Holding its front feet forward, it allows the current to spread out their strainer-like fringes of long hairs. These retain whatever food is swept against them, and the mouth of the nymph is conveniently near at hand. It uses its feet for standing but moves from place to place by means of swift strokes of its finely developed tail fin, supplemented by synchronous backward strokes of its strong tracheal gill covers. It has almost the agility and swiftness of a minnow.

---

*A "water-glass" is any vessel having opaque sides and a glass bottom, of convenient size for use. An ordinary galvanized water pail with its bottom replaced by a circular glass plate set nearly flush, is excellent.

2. *Ordinary Foragers.*—These are the members of lotic societies that lack such specialized means of gathering food from the passing current, and that forage by more ordinary methods. They live for the most part on the sides of stones and underneath them, and not on their upper surfaces. These also live where the water runs swiftly, and, for the most part, out of the reach of those fishes that invade the rapids. There are two principal categories among them: *a.* Free-living forms that are more or less flattened or limpet-shaped. *b.* Shelter-building forms, that are in shape of body more like the ordinary members of their respective groups.

a

The limpet-shaped forms are members of several orders of insects, worms and snails. Their flattened form and appressed edges are doubtless adaptations to life in currents. They adhere closely, and are on account of their form, less likely to be washed away; the current presses them against the substratum.

Not the most limpet-like but yet the best adapted for hanging on to bare stones in torrents is the curious larva of the net-veined midge, Blepharocera (see fig. 159 on p. 259), an inhabitant only of clear and rapid streams. The depressed body of this curious little animal is equipped with a row of half a dozen ventral suckers, each of which is capable of powerful and independent attachment to the stone. So important have these suckers become that the major divisions of the body conform to them and not to the original body segments. On these suckers, used as feet, the larva walks over the stones under the swiftest water, foraging in safety where no enemy may follow.

Most limpet-like in form of all is the larva of the Parnid beetle, Psephenus, commonly known as the "water-penny" (see fig. 160 on p. 260). It is nearly circular and very flat with flaring margins that fit down closely to the stone. It adheres closely and is easiest picked up by first slipping the edge of a knife under it. Viewed from above, it has little likeness to an ordinary beetle larvæ, but removed from the stone and over-turned, one sees under the shell a free head, a thorax with three short legs, an abdomen and some minute soft white segmentally arranged tracheal gills on each side.

Other insect larvæ that have taken on a more or less limpet-like form, are the nymphs of certain May-flies and of many stoneflies (fig. 111 on p. 204). The body is strongly depressed. The lateral margins of the head and thorax are extended to rest down on the supporting surface. The legs are broadened and are laid down flat so as to offer less resistance to the currents, and stout grappling claws are developed upon all the feet. Such is Heptagenia whose nymphs abound in every riffle and on every rocky shore.

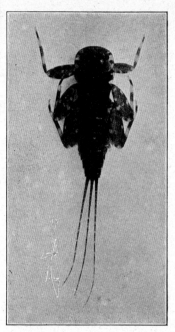

FIG. 218. The nymph of a may-fly (Heptagenia) from the rapids, showing depressed form of the body and legs.
(Photo by Anna H. Morgan.)

One may hardly lift a stone from swift water and invert and examine it without seeing them run with sidelong gait across its surface, outspread flat, and when at rest appearing as if engraven on the stone.

The head is so flat and flaring that the eyes appear dorsal in position instead of lateral as in pond-dwelling Mayfly nymphs.

A more remarkable form is the torrent-inhabiting nymph of Rithrogena whose gills are involved in the flattening process. They also are flattened and extended laterally and rest against the stone. But,

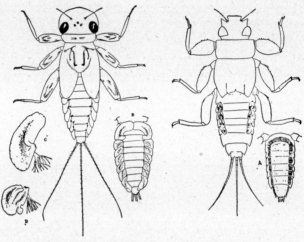

FIG. 219. Parallel development of limpet-like form of body in two mayflies. Right, the nymph of *Ephemerella doddsi;* A, the overturned abdomen. Left, the nymph of *Rithrogena mimus;* B, the overturned abdomen; C, the foremost gill; D, the second gill. (Courtesy of the Utah Agricultural Experiment Station).

most remarkable of all, the anterior pair is deflected forward and the posterior pair, backward, to meet on the median line beneath the body, and both are enlarged and margined; By the close overlapping of all the gills of the entire series there is formed a large oval attachment-disc of singularly limpet-like form.

A similar flat attachment-disc is formed on the ventral side of the mayfly nymph shown in figure 219, but on a wholly different plan. The gills are not involved in the disc, but instead the body itself is flattened and shaped to an oval form underneath, and fringed with close set hairs.

There is in the mayflies a rather close correlation between the degree of flattening of the body and the rate of flow of the water inhabited. It is well illustrated by the allies of Heptagenia; also by those of Ephemerella, among which occur swift-water forms. Epeorus, Iron and Rithrogena form an adaptive series. Among the Parnid beetles, Elmis (fig. 214b), Dryops and Psephenus (fig. 160) form a parallel series.

There are snails that dwell in the rapids. The most limpet-shaped of these is Ancylus (fig. 160 on page 260) whose widely open and flaring shell has in it only a suggestion of a spiral. Certain other snails (such as *Goniobasis livescens*) are of the ordinary form and are able to maintain themselves on the stones by means of a very stout muscular closely-adherent foot. Similarly, a number of flatworms, that adhere closely are found creeping in the rapids.

Shelter-building foragers are numerous in individuals but few in kinds. One tube-dweller, Hydropsyche, is a plancton gatherer and has been already discussed. There are other shelter building caddis-worms living among stones in running water. Ryacophila builds at close of larval life a barricade of stones as shown in the fig. 125 on page 217, and shuts itself in

FIG. 220. Two pupal cases of the caddis-fly, Ryacophila, removed from the stones.

and spins about itself a brownish parchment-like cocoon of the form shown in the accompanying figure. Helicopsyche constructs a spirally coiled case that is strikingly like a snail shell, and fastens it down closely in the shallow crevices of stones on exposed surfaces.

FIG. 221. The spirally coiled cases of the caddis-worm, Helicopsyche.

A number of other caddis-worms build portable cases of sand and stones. Those of Gœra (fig. 222) are heavily ballasted by means of stones attached at the sides with silk. These lie down flat against the bottom and doubtless serve the double purpose of deflecting the current and preventing the case from being washed away.

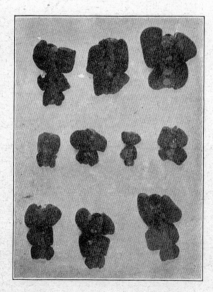

FIG. 222. Stone-ballasted cases of caddis-worms of the genus Gœra.

The tubes of the midges are here made of less soft and flocculent materials than in still waters. Tanytarsus makes an especially tough case of a pale brownish color, like dried grass. It is of tapering form, and easily recognized by the three stay lines that run out from the open forward end. A small greenish yellow larva with rather long antennæ lives within, and protrudes its pliant length in foraging on the algal herbage that grows about its front door. And there are many other lesser midges whose larvæ dwell in silt - covered tubes on rocks in the rapids. Often they occur so commonly as to almost cover the surface.

FIG. 223.   Larval cases of the midge, Tanytarsus, attached to a stone in running water.

Shelters also limpet-shaped—It should be noted in passing that this flattened form, which is characteristic of so many members of lotic society, is characteristic not only of the living animals but also of their shelters. The tarpaulin-like web of the moth *Elophila fulicalis* is flat, and the pupal shelter is quite limpet-shaped. The case of *Leptocerus ancylus* is widely cornucopia-shaped, its mouth fitted to the stone. The coiled case of Helicopsyche is a very broad spiral, closely

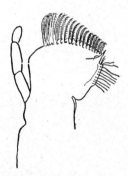

FIG. 224.   The maxilla of a mayfly, *Ameletus ludens*, showing diatom rake.

attached in the hollows of stones and crevices of rock ledges. The case of the caddis-worm, Ithytrichia, (fig. 162 on p. 262) is broadly depressed.

Thus the impress of environment is seen not only in the form of a living animal but also in that of the non-living shelter that it builds. In this there is a parallel of form in the secreted shell on the back of the snail, Ancylus, and manufactured shell on the back of the caddis-worm, Helicopsyche. One would have to search widely to find better examples of the effects of environment in molding to a common form these representatives of many groups of very diverse structural types. Two of them, at least, were sufficiently like lotic mollusca to have deceived their original describers. Psephenus was first described as a limpet and Helicopsyche as a snail.

Foraging habits—The food of the herbivores in lotic societies is algæ. There are none of the higher plants present, save a few mosses of rather local distribution. It is not surprising therefore that the food gathering apparatus of these forms should present special adaptative peculiarities. The mouth-parts of mayflies and of midges show much development of diatom rakes and scrapers. For scraping backward the labrum is often used. In the net-spinning caddis-worms it is bordered on either side by a stiff brush of bristles, and in midge

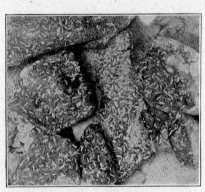

FIG. 225. The sheltering tubes of midge larvæ. Photographed under running water on the rocky bed of a stream.

larvæ there is developed both before and behind its border a considerable array of combs and rakers. In use the head is thrust forward, and these are dragged backward across the surface that supports the growth of diatoms and other algæ.

The principal carnivores of the rapids are the nymphs of stoneflies (see fig. 111 on p. 204) and a few small vertebrates. Among the latter are the insect-eating brook salamander, Spelerpes, and a number of small fishes, such as darters, dace and minnows.

THE TWO PRINCIPAL FISH ASSOCIATIONS
OF LAKE GEORGE, N. Y.

From the senior author's Report on a Biological Survey of Lake George
Courtesy of the New York State Conservation Commission

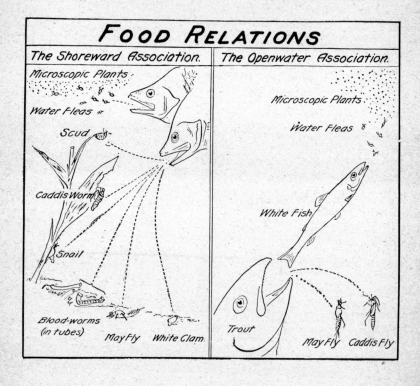

CHAPTER VII

INLAND WATER CULTURE

ABORIGINAL WATER CULTURE

HARDLY any native species found by the white man in America had done so much to alter and improve its environment as had the American beaver. Certainly the red man had done less. Thousands of acres of fertile valley land now tilled by American plowmen was levelled up behind beaver dams. These followed one another in close succession in the valley of many a woodland stream. The wash from the hills settled in their basins. As they were filled, dams were built higher, and thus the rich soil grew deeper.

The beaver was a builder of ponds. His only method was by damming gentle streams. He cut down trees with his great chisel-like teeth, trees often six, eight, or ten inches in diameter. He cut off their boughs and

377

drew them to the place where a dam was to be con-structed. He piled them as a framework for a dam, weighted them in position with stones, filled the inter-stices with trash and leafage and covered the water side over completely with mud, making it impervious. And when the water had risen behind it he built him a dome-shaped house on the edge of the pond thus created, having passageways opening beneath the water, and he plastered it over with mud. When marsh plants grew about the edges of the lands he had thus inundated, he cut channels through them for easy passage to his favorite feeding grounds. His staple food was the bark of aspens and birches that grew thickly near at hand, but this he varied with succulent shoots and tubers of aquatics. These nature planted for him, as soon as he had prepared his water-garden.

This was aboriginal water culture.

FIG. 226. An aboriginal water-garden. A beaver dam and pond. (From Morgan.)

WATER CROPS

FERTILITY dwells at the water side, where the essential conditions for growth—m o i s t u r e, warmth, air and light—abound. There Nature's crops are never failing. They are abundant crops compared with which the herbage of the uplands appear thin and scattering . If they are not our crops, that is not Nature's fault but our own. We have given all our toil and care to the cultivation of the products of the land, and have left the waters to produce what they might, often in the face of neglect and injury.

Time was when the waters furnished to man the most dependable part of his livelihood—fish and oysters and edible roots and excellent furs. That was before the days of agriculture. Primitive man, while gathering his fruit and roots and grains from the wild, saw the supply failing and planted a garden to increase his sustenance. Had he by like means endeavored to supplement his stores of water products, we might now have had a water culture, comparable with agriculture.

A number of native water plants furnished food to the red men in America. One of these, the wild rice

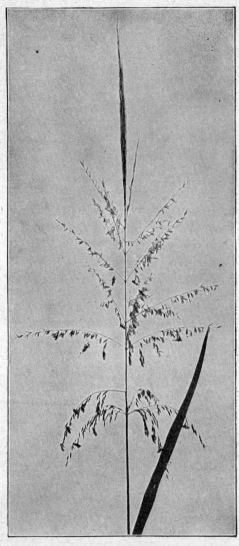

FIG. 227. A flower-cluster of wild rice, fertile above, staminate below. Little brown syrphus flies of the genus Platypeza cling to the staminate blossoms.

(fig. 227), is obtainable in our own markets in very limited quantity and at fancy prices: it grows as a wild plant still. The Indian ate both the nut-like seeds and the stocks of the wild lotus; also the tubers of the arrowhead, the stocks of the arrow-arum, the enormous rhizomes of the spatterdock, the succulent shoots of the cat-tail, and other rather coarse and watery wild plant products, that we esteem better food for muskrats than for men. The starch-filled tubers of the sago pondweed (fig. 228) are choice food for waterfowl, and if obtainable in sufficient quantity would probably be prized by men, for when cooked they are both pleasing in appearance and very palatable.

A number of rushes of different sorts were in aboriginal times used for coarse weaving of mats, etc.; and one of these, the narrow-leaved cat-tail, we have of late begun to use in new ways; in paper making and in

FIG. 228.   Tubers of the sago pondweed.
Potamogeton pectinatus.

cooperage.   The initial cut on the preceding page shows a field of cat-tail carefully cut and shocked for use in the calking of barrels that are to hold watery liquids.   The leaves are placed singly between the staves of the barrels, where they swell when wet, packing the joints tightly.

It may be that none of these plants will ever be cultivated.   Some are abundant enough for present needs without it. Wild rice is but another cereal grain, tho an excellent one. We already have garden roots in great variety of sorts that we prize more highly than we do these wild aquatics. The white water lily will be cultivated in the future for its beautiful flowers rather than for its edible tubers.

FIG. 229.   The white water lily, *Castalia odorata.*

The animal products of the water are more important. Aquatic molluscs, crustaceans, and vertebrates have ever furnished staple foods. Tho fresh water molluscs are no longer eaten, immense accumulations of their shells along some of our inland waterways bear silent testimony to the extent to which they were once consumed by the aborigines. Their shells also served other primeval uses, as cups and as scrapers. In our own day a new and important use has been found for them in the manufacture of pearl buttons and ornaments. They make the best of buttons, neat and durable and beautiful, a great improvement over the buttons of wood and metal formerly in use. The annual product of pearl buttons from this source is now worth many millions of dollars. It is all derived from wild

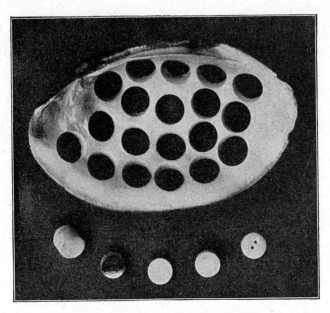

Fig. 230. Valve of a mussel shell, with "blanks" cut from it, in process of manufacture into pearl buttons.

mussels; the method in use is exploitation, not husbandry.

Fish—The great staple food product of the water is fish. In our day frogs are used but locally and fresh water crustaceans and other animals, hardly at all; but fishes are used everywhere. They have been a staple food from the beginning of human history, and probably will be to the end. Hence it is that inland water culture means to a large extent the raising of fishes.

*Fish culture** in America is in a very backward state as compared with animal husbandry in other lines. This is manifest in many ways; among them, the following:

1. There is lack of improved cultural varieties. Our fishes are wild fishes. Save for a few races of gold fishes—all fancier's fishes—and some not very desirable varieties of carp, hardly any improvements have as yet been made by selection and careful breeding.

2. There is lack of knowledge of the best kinds of forage for fishes and of how it may be provided for their use. This is half of the problem of raising any animal.

3. There is lack of any practical system of management, that provides for the breeding and feeding and rearing of stock, generation after generation, under control.

In what, then, does the fish culture of the present consist? Mainly in this one thing, the care of the young. This includes the gathering and hatching of fish eggs and the rearing of the young fishes thro their earlier stages on artificial food in hatcheries. By this means the enormous losses that occur under natural conditions in early life are avoided, and vast numbers of fry and fingerlings are grown to a size suitable for planting in natural waters. Thus far the methods are well worked out. Thus far our fish culture is brilliantly successful. But this is really only the first step. How these little fishes when turned loose in pond and stream shall find for themselves the means of a livelihood is the unsolved part of the problem. Planted here they seem to thrive; there, they fail. Every

*The substance of the following pages covering this subject was published by the senior author in the *Indianapolis News* in 1909, and again in the *Farmers Magazine* in 1912.

planting in a new place is more or less an experiment. Sheep culture would be in a state quite comparable with the fish culture of to-day, if after rearing lambs on the bottle they were turned loose in an unexplored forest to shift for themselves.

The hatcheries are raising fry and not fishes. This is, of course, what they were commissioned to do, the underlying idea being merely that of putting back into the lakes and streams a copious supply of young fishes to occupy the place of the adult fishes taken out. But experience has shown that the mere planting of fry soon reaches its effective limit, after which the planting of more fry is sheer waste. The conditions in the wild are not such as yield much advantage from this intensive propagation of the young. Oftentimes the fry planted in the trout streams about Ithaca may be found shortly afterward in the stomachs of the few adult trout that live in the same streams. Feeding fishes on the young of their own kind is not good husbandry.

The planting of fry and of fingerlings is effective where conditions permit of their growth. The removal of enemies is a supplemental measure of great value where practicable. The care of natural feeding grounds to prevent their destruction is very important, but usually impossible, for want of enlightened public opinion. Protecting of breeding fishes when on their spawning grounds—the time when they are most easily discovered and destroyed—is also very important. And the bringing back into habitable places of young fishes stranded in the side pools of bottomland streams, where they would perish with the evaporation of the water, is rescue work of a good sort. All these things are done in the interests of public fishing at the present day. They are such measures as are taken to preserve wild game in a forest or livestock on an open range. They have to do rather with hunting than with husbandry.

The day is coming—is already at hand—when he who wants fishes fresh from the water will have to raise them. Public waters are "fished-out." In spite of closed seasons, and frequent plantings of hatchery-reared fry, they continue to be "fished-out." With the growth of our population they are going to be always "fished-out;" and there is no hope for the future of any fishing that shall be worth while except in waters that are privately controlled.

This does not mean that there will be no fishing in the future. It only means that fish raising is going the way wild pig raising has gone.

When game began to fail—venison, wild turkeys, etc., the pioneer began to raise pigs. At first he gave them little attention, except at killing time, and furnished them no food. He raised them about as we raise fishes now. He turned them loose in the woods to forage for themselves as we now plant fish fry in the streams. They ranged the whole area where their food grew.

Nowadays, thousands of hogs are raised where one was raised then, but they do not run the range; they are kept in small lots, and the broad areas are devoted to raising forage for them. The present day method of obtaining our meat supply is very unromantic as compared with chasing a razorback hog with a shotgun through the woods at the end of the acorn season, but it is the inevitable way of progress in animal husbandry.

Raising animals and their forage together is not good husbandry. It is exceedingly wasteful and unproductive; yet that is the way we still raise fish in America. We ought to be doing better than this. It is idle to plant more fish in the water until we can supply more stuff for them to eat. And we cannot expect more forage to grow unless we provide suitable conditions.

When we raise other stock-feed we find a few perfectly definite things to be done:

1. We clear a field and prepare it.

2. We fence it to keep out enemies and undesirable competitors.

3. We plant it with selected seed; and after a period of growth,

4. We use the crop at the time of its maximum value.

All these things we shall have to do if we ever have a real fish culture. The first two of these things are usually cared for in the construction of fish-ponds; the other two are generally neglected.

The forage problem is less simple than is the raising of pigs on clover, for at least two reasons:

1. Plant foods are not eaten directly by the more valuable fishes, and often there are a number of turns-over of the food stuffs before the fishes are reached. For example, diatoms and other synthetic plancton organisms are eaten by water-fleas and midge larvæ, that are in turn eaten by little fishes, that are eaten by big fishes. There must be at least two turns-over— one kind each of plant and animal forage—since the desirable food-fishes are carnivorous.

2. There may be one or more changes of diet during development. Thus the pike when newly hatched eats such water-fleas as Simocephalus, (see fig. 92 on p. 186) picking them one by one with automatic regularly-timed snappings of its jaws. When grown a little larger it eats midge larvæ, mayfly nymphs and other small insects. Still later, it eats large insects and mixes small fishes in its diet; and as it attains full stature it restricts its diet to frogs and larger fishes. When grown it takes hardly anything smaller than a golden shiner.

Studies of the food of the common sunfish, *Eupomotis
gibbosus*, by the senior author ('08) have shown that in
Old Forge Pond, when one inch in length the food is
predominantly entomostraca and very small midge
larvæ.   When two inches in length, it is entomostraca
and midge larvae of larger size, together with small mayfly nymphs (Cænis) and minute snails.   When three
inches in length, it is grown midge larvæ, mayfly nymphs
and caddis-worms.   At this size apparently the diet of

Fig. 231.   The common sunfish.   *Eupomotis gibbosus.*
(Photo by George C. Embody)

entomostraca and small midge larvæ is outgrown, and
the fishes are seeking bigger game.

At three inches in length, this fish is itself the
favorite food of adult bullheads.

Excepting for a few fishes that range the open waters,
such as white-fish and lake herring, and that continue
to feed largely on plancton, there is at least one necessary shift of diet accompanying growth; that from
plancton to the food of the adult.   In an earlier chapter
(see p. 235) we have briefly indicated the principal
changes of diet then occurring.

The food relations of aquatic organisms are exceedingly complex. They change with age and season and situation. The eater and the thing eaten often exchange roles. Yet there are some fairly constant food dependencies between the major groups of organisms. These have been set forth by that veteran student of the forage problem, Prof. S. A. Forbes, in the table copied herewith (fig. 233), and this table indicates (what detailed food studies at large abundantly confirm) that fishes eat almost every living thing that the water offers.

FIG. 232. The nymph of the dragonfly, *Anax junius*, devouring a small sunfish.

The young of all fishes eat plancton. This sounds like one point of general agreement, until we reflect on the variety of organisms of which plancton is composed. Which of these are best for use in fish culture we scarcely know at all. Fortunately, they are of nearly universal distribution in shoal fresh waters, where the young of fishes are found.

Staple foods—While a list of all foods, eaten by all fishes would include practically every thing that is found in the water, yet when careful food studies are made there are a number of organisms so constantly recurring that they stand out as of prime importance. A few aquatic herbivores are found as commonly and as regularly in the stomachs of wild food fishes, as grass would be found in the stomachs of wild cattle. And just as stock feeding has made progress with the isolation and study and increase of the grasses, so fish culture would be advanced by study and cultivation of the staples of wild fish food.

PRINCIPAL FOOD RELATIONS OF AQUATIC ORGANISMS (ILLINOIS)	BACTERIA	ALGAE	HIGHER PLANTS	PROTOZOA	ROTIFERS	ENTOMOSTRACA	WORMS	CRAWFISHES	INSECTS	MOLLUSKS	FISHES	FROGS;TADPOLES	TURTLES	SERPENTS	BIRDS	MAN
TERRESTRIAL WASTES	X	X	X	X	X	X	X	X	X	X	X	X				
BACTERIA				X	X	X	X		X	X	X					
ALGAE				X	X	X	X		X	X	X	X				
HIGHER PLANTS								X	X		X		X		X	
PROTOZOA				X	X	X	X		X	X						
ROTIFERS				X		X			X	X						
ENTOMOSTRACA		X		X	X	X		X					X			
WORMS								X	X		X	X				
CRAWFISHES								X			X	X		X	X	
INSECTS		X						X	X		X	X		X	X	
MOLLUSKS							X	X			X		X		X	
FISHES						X	X	X	X		X	X	X	X	X	X
FROGS											X	X	X	X	X	X
TURTLES							X									X
SERPENTS															X	
BIRDS													X			X

FIG. 233. Forbes' (14) table of food (at left) and feeding organisms (above).

Our best fishes are carnivores, and the animals they eat are chiefly a few hardy, prolific, and widely distributed herbivores, such as water-fleas, scuds, midge larvæ, mayfly nymphs and other fishes. These feed, of course, on plants; but we hardly know as yet what plants are of most value to them. They thrive where herbage abounds; and yet we know that abundance of herbage may not necessarily mean good crops; for weeds may be much more conspicuous in a pasture than the close-cropped grasses that yield the forage there. Certain species of pondweeds have been shown by Miss Moore ('15) to be often used as green food, and Birge ('96) has given many notes on the food preferences of herbivorous plancton crustacea.

The above mentioned staples invite much attention but we shall have space for noticing but a few representatives of the groups to which they severally belong.

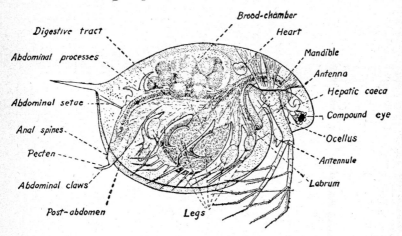

FIG. 234. Daphne (after Dodds).

Water-fleas—As a typical representative of this great group of herbivores, we may speak of Daphne (fig. 234). Its manner of life and its enormous reproductive capacity have already been briefly mentioned (pp. 186–7 and 306). It is a very valuable animal in water culture on account of its ability to turn the great growths of colonial diatoms and algæ into excellent food for fishes. Little is known, as yet, unfortunately, about the conditions that make for its growth. Plancton studies of water-fleas have consisted in the main of the counting of individuals in random catches; and, as Hæckel ('90) long ago pointed out, this has about as much economic value as the counting of straws would have in an oat field.

The extraordinary growths of certain plancton algæ (Anabæna, Aphanizomenon, etc.) that often give trouble in water-supply reservoirs, might be made into fish food through the agency of daphnias, if we only had learned how to manage our water crops.

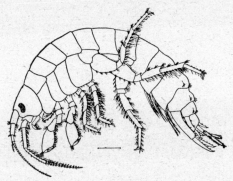

FIG. 235. Gammarus fasciatus (after Paulmier).

Water-fleas are of very great value as food for young fishes, they form also a considerable part of the food of such larger fishes as are equipped with gill strainers for gathering them out of the water. They are, of course, largely absent from the water during the winter season. Their value as forage organisms lies in their good quality and their extraordinary reproductive capacity.

The scuds—This group of herbivores is typified by Gammarus (fig. 235) a hardy, wide-ranging habitant of the water weeds. It swims well, yet prefers to occupy the sheltering crevices of dense leafage. It can leap and dodge like a rabbit. It feeds on a great variety of both living and dead herbage. It is itself a favorite food for most fishes.*

The scuds are easily managed in pond culture. They are not remarkably prolific. As already mentioned on page 190, the possible progency of a single pair in one year is somewhat less than 25,000. But they carry their young in a pectoral brood pouch until well equipped for life.

The chief merits of the scuds as forage organisms (in addition to desirability as food) lie in their hardiness, their ability to find a living and to take care of their own young until well started in life, their constant succession of overlapping broods thro the season and their permanent residence in the water.

There are other herbivorous crustaceans of somewhat similar habits, among which the fresh-water prawn, Palæmonetes, is probably useful as fish forage.

Midge larvæ—Larvæ of midges of the genus Chironomus popularly known as "blood-worms" (fig. 236) are

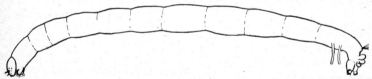

FIG. 236. A "blood-worm."

of prime importance as fish food. Small ones are eaten almost as universally as are plancton entomostraca, and the large ones continue to be eaten whenever obtainable by fishes as large as adult trout and white-

---

*Its value has long been recognized by fishermen; on account of its abundance in an excellent trout stream at Caledonia, N. Y., it has been locally known as the "Caledonia shrimp."

fish.   In a most extensive examination of the contents of fish stomachs Forbes ('88) found them "of remarkable importance, making in fact nearly one-tenth of the food of all the fishes studied."   Ferguson fed some red-bellied minnows (*Chrosomus erythrogaster*) for 22 days all the midge larvæ (*Chronomus viridicollis*) they would eat and nothing else.   The grown minnows ate on an average twenty-five blood-worms per day; the half-grown ones, eleven.   The senior author ('03) found that 25 brook trout taken at random from one of the best natural ponds of the New York State Fish and Game Commission at Saranac Inn, N. Y., had in their stomachs more than 100 blood-worms each.

Midge larvæ are among the most ubiquitous of freshwater organisms.   They feed mainly upon diatoms, and other simple organisms found in water or growing sessile on or round about their homes; the larger ones eat also the disintegrating tissues of the higher plants.   They dwell among all sorts of aquatic plants, spreading their thin filmy tubes in every crevice or along the stems.   Little is seen of them there on casual observation.   They are like the rodents of the fields, hidden in their runways.   But one cannot place a handful of any water weed in a dish of water without soon seeing some dislodged midge larvæ swimming about the edges with characteristic figure-of-8-shaped loopings of the body.

They dwell on the bottom (see fig. 134 on p. 226). Indeed, as already noted, they may dwell far out on the bottom under the deep water of great lakes.   Here in deep darkness and heavy pressure they dwell in enormous numbers feeding upon the rich spoils of the plancton rained down on them by gravity from above.   They often fill the soft bed with their silt-covered flocculent tubes.

These tubes, like the ones on the stems, open to the surface at both ends. The larva, within, holding on to the silken lining of the walls with its claws, swings its body in vigorous undulations, driving a current of water thro the tube. This serves for respiration. It also serves to drive diatoms and other food organisms into net-like barriers spun across the exit; these barriers are repaired or renewed after every catch. Food is thus carried into the shelter of the case. But food is also gathered from exposed surfaces whenever it can be reached from open ends of the tube. It is gathered by scraping the sessile diatoms and algæ from stems. For such work the mouth of the larva is equipped with elaborate rakes and scrapers.

The larva of Chironomus is relatively simple. It appears much less complex in organization than are many of its insect competitors. It has a cylindric worm-like pale and naked body with a bifid proleg underneath at the front and a pair of prolegs behind, caudal tufts of bristles, and a few simple gills. The prolegs are armed with hooks and on them it creeps somewhat like a looping caterpillar. From its mouth it spins the fluid silk, and spreads it ere it hardens with the front proleg. All in all, it is a shy and defenseless and secretive creature, without any special gift of locomotion.

This apparent weakling has been able to possess itself of the entire littoral region of the earth, perhaps by reason of the following characteristics:

1. Ability to live on foodstuffs that have a very general distribution.
2. Ability to build its own shelter.
3. Consequent adaptability to variety of conditions.
4. Great reproductive capacity.
5. Brief life cycle.

Chironomus lays several hundred eggs, and in the warm season a generation may completely develop in five or six weeks; so the very considerable increase of one brood may be rapidly repeated in geometric ratio.

The limitations to its use as a forage organism in fish ponds lie in its complicated life history. It quits the water at the end of the pupal stage. It flies away, mates in the air, and returns to the water to lay its eggs. During its aerial life it is not easily managed.

Mayfly nymphs constitute one of the most important groups of aquatic herbivores. We single out Callibætis for illustration of another staple fish food. It is an active nymph that swims from place to place by means

FIG. 237. The nymph of Callibætis: Drawing by Anna H. Morgan.
(From *Annals Ent. Soc. America*)

of quick strokes of its tail and gills, and that clambers freely about over shore vegetation. It is an artful dodger; and it is protectively colored. It feeds on a great variety of vegetable substances living and dead, and hence finds abundant food in every weedy pond. It is eaten by every carnivore in the pond that can catch it; and doubtless it has many enemies that exceed it in swiftness and many others that lie in ambush and capture it by stealth. Hence, tho nearly always present, it rarely appears very abundantly in old ponds.

The life cycle of Callibætis is run in less than six weeks. A single female may lay 1000 eggs. If all these were to develop and reproduce, the increase from a single pair during one summer season would be something like this:

1st brood 1,000 (half females)
2d brood 500,000.
3d brood 250,000,000.
4th brood 125,000,000,000.

These alluring possibilities of increase in an organism that is choice fish food once led the senior author into a series of experiments that extended through two years and that met with uniform failure because the breeding of the mayflies could not be controlled. The rearing was easily managed but even with the largest measure of freedom that could be provided, the adults would not mate and lay eggs in captivity. The problem of their successful artificial propagation is still unsolved. However, there has never been a new pond opened at the Cornell University Biological field station, that has not received the eggs of wild females of Callibætis, and that has not raised a good crop of the nymphs ere their slower-breeding carnivorous enemies developed.

Mayflies, like Callibætis and the little Cænis, that have a number of broods each season with overlap of generations, are suited for use in forage propagation because at all times of the year nymphs of good size are present in the water. On the other hand, such forms as *Blasturus cupidus*, which flies in May, and *Siphlonurus alternatus* which flies in June, are absent from the water at the close of their breeding season or are represented there only by eggs and very minute nymphs.

Best known of the mayflies that fishes eat are the nymphs of the big burrowing Hexagenias from lake and river beds. Food examinations have abundantly shown their importance. However, they develop slowly, requiring at least two years to reach maturity.

The Hexagenia nymphs are natural associates of bloodworms on the lake bottom. They, and the blood-worms with them, and the entomostraca swimming above them are the mainstay and dependence of the lake's fish population.

Other herbivorous insects of promise as forage organisms are caddis-worms and aquatic caterpillars. Other invertebrates are a number of pond snails. But the animals above discussed we regard as most important.

Forage fishes—The largest single item in the bill-of-fare of fishes generally is other smaller fishes. Herbi-

FIG. 238. The golden shiner.
(*Photo by George C. Embody*)

vorous fishes, non-competitors for food, may therefore be used to furnish a principal crop of animal forage. For this use carp are objectionable because they grow too fast and soon become too large to be swallowed by the other fishes. They eat the eggs of bass, and root up the bottom and tend to exterminate their own vegetable forage. Minnows are also objectionable because they eat the eggs of other fishes. But very valuable for such use are the golden shiner (fig. 238), and the gizzard shad, (*Dorosoma cepedianum*), of our great rivers. Even the goldfish is an excellent agent for turning masses of blanket algæ and other soft fresh vegetable foods into excellent forage for larger fishes.

The way of economic progress—The future of fish culture lies in further scientific studies to be made along the lines that have proven of value in the raising of land animals. More knowledge is what is needed:

1. Intimate detailed knowledge of the fishes themselves is needed; knowledge of their natural history, their requirements of food and of protection for their young; their enemies, internal and external; their natural races and possibilities of improvement by breeding. Only such knowledge can furnish a basis for developing methods of control.

2. Equally detailed knowledge is needed of the economic species that furnish forage or that menace the welfare of the cultivated species; knowledge of all the more important ones, from the forage fishes, crustaceans, insects, snails, etc., even down to the diatoms. The product must be followed to its principal sources and the cultural relations that all these organisms bear to each other must be better understood. The enemies of every stage of fish life must be studied (fig. 239).

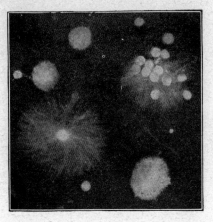

FIG. 239. Eggs of the pike, *Esox lucius*, overgrown with two species of fungus.

3. More knowledge is needed of the water bodies themselves; knowledge of their physical, chemical and hydrographic conditions, their purity, contamination, and all other conditions that affect the welfare, that promote or hinder the normal growth and activities of the useful organisms contained in them. We must know these things in order to know how to make and keep the waters productive.

Knowledge is being accumulated in all these lines in a slow and desultory way, thro the voluntary activity of many diverse and widely scattered agencies. Fish culture has not yet had the benefit of that efficient agency of economic progress that has brought such rapid improvement in animal husbandry—the experiment station. A fish cultural experiment station is what is now urgently needed: an institution equipped for water culture, and charged with the duty of carrying out a well planned line of experiments bearing on its economic problems. This is needed to supplement the hatcheries and to bring their work to fruition.

WATER CULTURE AND CIVIC IMPROVEMENT

THE three chief interests of the public in water culture lie (1) in making the waters productive; (2) in keeping the waters clean and (3) in preserving the beauty of the waterside. Happily, these are concordant, and not conflicting interests.

Another interest of everybody is in pure water to drink. For city-dwellers, public water supplies must be kept uncontaminated—a matter of ever increasing difficulty as our population grows. This vast subject falls without our present scope: its literature may be found by following up a few references (Whipple, *et. al.*) given in the bibliography at the close of this volume.

There are two very large reclamation enterprises, with which water culture should have much to do in the future:

1. The reclamation of waste wet lands, and
2. The utilization of water reservoirs.

A few words may be said here concerning each of these.

WHAT SHALL BE DONE WITH THE MARSHES?

There are millions of acres of waste wet lands in America, that are producing little or nothing of value. That this land will yet be made to contribute much more largely to human sustenance, there can be no doubt: for,

1. It is the richest of all the land, in foodstuffs that make for soil fertility. It contains organic remains accumulated for ages, together with the wash from surrounding slopes.

2. It is generally the best located of all the land with respect to transportation facilities. Inland marshes almost everywhere are traversed by railways, their levels having invited the attention of the route-locating engineer; many marshes border on navigable waterways.

3. It is the last of the land available for occupation, and with our population quadrupling every century, the pressure for room is becoming ever more intense.

While it is inevitable that most of this land will yet be used for production of human food, it is by no means certain how this may best be done. Drainage is the one method hitherto tried, but drainage has its serious limitations:

1. Much of the wet land cannot be profitably drained.

2. Its value as a water reservoir is largely destroyed by drainage.

There is another plan for making marshes productive that has not yet been tried on any adequate scale—a plan that involves water culture as well as agriculture. The marshes—now neither wet nor dry—cannot be used as they are; but if by a shifting of some of their topsoil they be made in part into permanently dry, and in part into deeper reservoirs of water, they might then be cultivated in their entirety. The dry part would be available for ordinary agricultural use and crops can be grown by methods already well worked out. The permanent water could be made to produce fish and fish forage and other water crops. The advantages of this plan over drainage would appear to be the following:

1. Increased productiveness.
2. Permanent water storage.
3. Diversifying of crops: it would not be merely adding more of crops already extensively cultivated.
4. Diversifying the industries of the people.
5. Completer utilization of the wet areas.

THE WASTAGE OF RESERVOIR SITES

There is another service that water culture may render to great public works. It may make water reservoirs productive. The various measures now being widely considered for the development of our water resources should be co-operative rather than conflicting. The making of reservoirs for holding the surplus rainfall near the headwaters of streams, allowing it to flow as needed, should result in three distinct and permanent civic benefits:

1. Permanent water power.
2. Continuous navigation.
3. Increased production of food.

One of the things that has stood in the way of the development of reservoirs has been the necessity for condemnation of valuable agricultural lands needed for the reservoir site. Such lands when covered with water, are of course, removed from agricultural use. But they might yet be used for water culture, and indeed the value of the resulting crops might thereby be increased.

Some special development of the water bed would, of course, be needed to fit them for an intensive water culture. The one great open basin of water, now full and now reduced, that is the usual thing in reservoirs, would hardly suffice. But with no extraordinary increase of cost the greater part of the bottom, especially in shoal water, might be divided into fish ponds, so constructed as to be under control. By deepening these considerably and using the excavated earth for building strips of dry land between them, the holding capacity of the reservoir might be increased. It would be increased by just so much as the volume of earth taken from below and placed above the high water level. Then as much water as under the present plan could be drawn off for power or navigation, and the residue in the pond bottom would suffice for maintenance of the fishes therein.

On this plan, in a reservoir of 100 acres having 90 acres of shoal water on which fish ponds could be developed, 50 acres could be permanently devoted to fish raising, and at least half or much more to agricultural crops, without interfering with its efficiency for water storage and regulation of stream-flow. This would be much better than having it all lie fallow to the end of time. It would transform a water waste into a water garden. Incidentally, it would cure also the unsightliness of a vast area of exposed and reeking mud during the season of low water.

The beauty of the shore-line—Another public interest with which water culture must ever be identified is that of preserving the beauty of the landscape. As nature has given of her bounty to the waterside, so also she has lavished her beauty there.

What flowers adorn the shore-line! The fragrant water lily, the stately lotus, the queenly iris, the brilliant hibiscus, the soft blue pickerel-weed, the sweet forget-me-not! What foliage in pondweed and water

FIG. 240.   The common wild forget-me-not.

shamrock, in arrowhead and arrow-arum, in water-shield and spatterdock! What exquisite submerged meadows the pondweeds, bladderworts and the milfoils make! How inviting are the shores where these abound, how unattractive, those from which these have been removed.

The landscape belongs to all.   Its condition affects the public weal.   It is good to dwell in a place where the environment breeds contentment; where peace and plenty and satisfaction grow out of the right use of nature's resources; where wise measures are taken to preserve the bounteous gifts of nature and to leave them unimpaired for the use and benefit of coming generations.

Much of the scenic beauty of every land lies in its shore lines; and it should be a part of public policy to keep unimpaired as far as possible the attractiveness of all public waters.   Streams differ far less from one another in their own intrinsic characters than in the way they have been used by the hand of man.   They differ less by topography and latitude; far more by the cleanness of their waters, by the trees that crown their headlands, and by the flower-decked water-meadows that fill their bays and shoals.   The famous distant lakes and streams that attract so many people far from home every summer are not more beautiful or restful than many homeland waters once were, or might again be, were but a little public care exercised to keep their waters clean and the beauty of their shores and bordering vegetation unspoiled.

Private water culture—Great as are the benefits to be hoped for in public works, those to be derived from the application of a rational water culture to private grounds are probably in the aggregate far greater.   On thousands of farms there are waterside waste lands, lying bare and abused, that might be reclaimed to usefulness and beauty through intelligent water culture.

The making of a pond on the home farm is good work for the slack season; and once properly constructed it is permanent, and will with a minimum of attention yield returns out of all proportion to its cost. It will yield fresh fish for the table. It will yield healthful sports for the boys and girls who should be kept at home; angling, and swimming in the summer and skating in the winter. It will yield beauty; the beauty of a mirroring surface, reflecting trees and hills and

FIG. 241. A beautiful cover for a mud bank. The water-shamrock, Marsilea, in front, then arrowheads, then sedges.

sky and passing cloud; the beauty of the aquatics planted on the shore line: the beauty of the water animals, of flashing dragonfly and gyrating beetles, and leaping fishes. It will add to the joy of living.

The accompanying diagram is intended as a suggestion for the development of a tract of upland waste wet land into a water garden. Its noteworthy features are found in the provision for growing forage under control, and, in so far as need be, apart from the animals

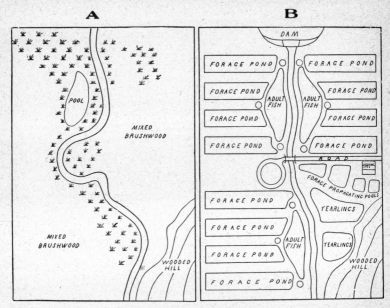

FIG. 242. Diagram illustrating the conditions for fish production on an 80 acre tract of wet upland, traversed by a trout stream. A, in a wild state. B, equipped for intensive fish raising.

Area devoted to fish, in *A*, one acre more or less; in *B*, one acre of enclosed ponds.
Devoted to fish forage, in *A* the same acre of open stream; in *B*, forty acres of ponds, planted and under control.
Devoted to land crops, in *A* none,—it is all too wet and sour; in *B*, all the made land between the ponds.

that are to eat it.   This is a suggestion for the application of the principles discussed in the earlier pages of this chapter. There is, of course, nothing original about it: it is what has made modern animal husbandry possible. It has not been applied to fish culture, however, and we are not able to give any figures of production because it has not been tried out in a practical way even on such a scale as is here shown.

Swamp Reservations—Now, having presented a plan for complete utilization of the marshes, we hasten to add that we believe it would be a great misfortune if

all the marshes were to be "improved." Some of them
are already serving their best use as refuges and breeding grounds of wild water fowl. In all of them there is
a whole wonderful fauna and flora that we could ill
afford to lose. That these would be lost under an

Fig. 243. Wall painting
from an ancient Egyptian
tomb showing the plan of a
house with a water-garden.
(After Brinton).

intensive water culture is highly probable (see fig. 244),
for our own cultivated crops are in the main successful
about in proportion as we eliminate the wild to make
room for them.

Since the wet land is almost the last of the unoccupied
land remaining near to the centers of human habitation,
and since it is the dwelling place of the largest remnant
of native wild life, we should not be taking measures for

410 Inland Water Culture

FIG. 244. A pond at Lake Forest, Ill., containing islands covered by butt
For effects of grazing, c

making it over to cultural uses without at the same time providing reservations where the wild species may be preserved for future generations. Each of these wild species is the end product of the evolution of the ages. When once lost it is gone forever: it can never be restored. We are not wise enough, nor farsighted enough to know whether the qualities lost with it would ever be of use to our posterity. We are now only at the beginning of knowledge of our plant and animal resources.

But quite apart from any possible economic values that these creatures of the wild may possess, they have other values for us that we should not ignore. Ere

ısh and divided by a pasture fence. The left hand end is closely pastured.
ıre the extreme ends.

their destruction is complete, public reservations should be made to preserve the best located of the marshes for educational uses. As we have need of fields and stock-pens because we must be fed, so also we have need of this wild life because we must be educated. It was with our forefathers in their early struggles to establish themselves in the New World: it conditioned their activities, lending them succor or making them trouble. In its absence it will be harder to comprehend their work. The youth of the future has a right to know what the native life of his native land was like. It will help to educate him.

Exploitation is reaping where one has not sown. Mere exploitation is but robbing the earth of her treasures. Usually it enriches only the robber, and him but indifferently. Getting something for nothing usually does not pay. It tends to rob posterity.

Exploitation is the method of a bygone barbarous age—an age when men, emerging from savagery, acquire dominion over earth's creatures ere attaining to a sense of responsibility for their welfare.

Conservation is the method of the future. It means greater dominion and completer use, but it also means restraint and regard for the needs of future generations. We are urging that in the use of our aquatic resources, the wasteful methods of exploitation be abandoned; and in two directions:

1. We urge that water areas, adequate to our future needs for study and experiment, be set apart as reservations and forever kept free from the depredations of the exploiter, and of the engineer.

2. We urge that in those areas which are to be made to contribute to human sustenance, the wasteful, destructive and irresponsible practices of the hunter be abandoned for the more fruitful and fore-looking methods of the husbandman.

SELECTED BIBLIOGRAPHY

Adams, Chas. C., and others. 1909. An ecological survey of Isle Royale, Lake Superior. pp. 468. Biol. Surv. Mich. Rept.

Adams, C. C. and Hankinson, T. L. 1928. The ecology and economics of Oneida Lake Fish. N. Y. State Coll. of Forestry, Bull. 1:247–547.

Aldrich, J. M. 1912. The biology of some western species of the Dipterous genus Ephydra. N. Y. Ent. Soc. Jour. 20:77–99. 3 pls.

Alexander, C. P. 1920. The craneflies of New York. Part II. Biology and phylogeny. Cornell Univ. Exp. Sta. Mem. 38:695–1042. 85 pls.

Allen, A. A. 1914. The red-winged blackbird: a study in the ecology of a cat-tail marsh. Linnæan Soc. New York Proc. pp. 43–128. 22 pls.

Allen, T. F. 1888–94. The Characeæ of America. New York.

Allen, R. W. 1914. The food and feeding habits of freshwater mussels. Biol. Bull. 27:127–144. 2 pls.

Alm, G. 1929. Prinzipien der quantitativen Bodenfaunistik und ihre Bedeutung fur die Fischerei. Internat. Verein f. Limnol. pp. 168–180.

Baker, F. C. 1910. The ecology of Skokie Marsh with particular reference to mollusca. Ill. State Lab. Nat. Hist. Bull. 8:441–497.

Behning, A. L. 1924. Zur Erforschung der am Flussboden der Wolga lebenden Organismen. Wolga-Station Monogr. 1:1–398, Maps.

Birge, E. A. 1895–6. Plankton studies on Lake Mendota. Wisc. Acad. Sci. Trans. 10:421–484: 5 pls., and 11:274–448. 27 pls.

Birge, E. A. 1916. The work of the wind in warming a lake. Wisc. Acad. Sci. Trans. 18:341–391.

Birge, E. A. 1918. A second report on limnological apparatus. Wisc. Acad. Sci. Trans. 20:533–552.

Birge, E. A. and Juday, C. 1909. A summer resting stage in the development of Cyclops bicuspidatus. Wisc. Acad. Sci. Trans. 16:1–9.

Birge, E. A. and Juday, C. 1911–1914. The inland lakes of Wisconsin. 1911. The dissolved gases of the water and their biological significance. Wisc. Geol. and Natl. Hist. Survey, Bull. 22. 1914. Hydrography and Morphometry. *Ibid.* Bull. 27.

Birge, E. A. and Juday, C. 1914. Limnological Studies of the Finger Lakes of New York. U. S. Bur. Fish. Bull. 32:525–609 (with maps).

Birge, E. A. and Juday, C. 1921. Further Limnological Observations on the Finger Lakes of New York. U. S. Bur. Fish. Bull. 37:211–252.

Birge, E. A. and Juday, C. 1920. A Limnological Reconnoissance of West Okoboji Lake (Iowa). Univ. of Iowa, Nat. Hist. Studies, 19:1–56.

Birge, E. A. and Juday, C. 1922. The inland lakes of Wisconsin: the Plankton, its quantity and chemical composition. Wisc. Geol. and Nat. Hist. Surv. Rept. 64:1–222.

Brauer, A. 1909. Die Süsswasserfauna Deutschlands. 19 vols. By 32 authors.

Carpenter, K. E. 1928. Life in inland waters, with especial reference to animals. London 267 pp.

Carpenter, W. B. 1891. The microscope and its revelations. 7th edition. pp. 1099. Phila.

Clemens, W. A. 1917. An ecological study of the mayfly Chirotenetes. Univ. of Toronto Studies. 17:1–43, 3 pls.

Clemens, W. A. 1923. The Limnology of Lake Nepigon. Univ. of Toronto Studies. 11:1–188.

413

Coker, R. E. 1914. Water-power development in relation to fishes and mussels of the Mississippi. U. S. Bureau Fish. Document No. 805. pp. 28. pls. 6.

Coker, R. E. 1915. Water conservation, fisheries and food supply. Popular Sci. Monthly, 36:90–99.

Coker, R. E. and others. 1919. The natural history and propagation of fresh-water mussels. U. S. Bur. Fish. Bull. 37:75–181.

Collins, F. S. 1909. The green algæ of North America. Tufts College Studies, Vol. II, No. 3. (Sci. Series.)

Comstock, A. B. 1911. Handbook of nature-study. Ithaca. pp. 938.

Comstock, J. H. 1925. An introduction to entomology. Ithaca. pp. 1044.

Conn and Washburn. 1908. The algæ of the fresh waters of Connecticut. Conn. Geol. and Nat. Hist. Survey Bull. No. 10. pp. 78. pls. 44.

Conn, H. W. 1905. The protozoa of the fresh waters of Connecticut. Conn. Geol. and Nat. Hist. Survey Bull. No. 12. pp. 69. pls. 34.

Cowles, H. C. 1901. The plant societies of Chicago and vicinity. Bull. II. Geog. Soc. Chicago. Also Bot. Gaz. 31:73–108, 145–182.

Dachnowski, A. 1912. Peat deposits of Ohio. Geol. Surv. Ohio Bull. (4) vol. 16.

Dakin, W. J. Aquatic animals and their environment. Internat. Rev. Hydrobiol. 5:53–80.

Darwin, C. 1875. Insectivorous plants. London.

Dodd, G. S. 1915. A key to the Entomostraca of Colorado. Univ. of Colorado, Bull. 15:265–298.

Dyche, L. L. 1910–1914. Ponds, pond fish, and pond fish culture. State Dept. Fish and Game. Kansas. Part I on ponds, 1910. Part II on pond fish, 1911. Part III on pond fish culture, 1914.

Ehrenberg, C. G. 1838. Die Infusionstierchen als vollkommene Organismen. Leipzig.

Embody, G. C. 1912. A preliminary study of the distribution, food and reproductive capacity of some fresh-water Amphipods. Internat. Revue der gesamten Hydrogiologie und Hydrographie. Biol. Suppl., III Serie.

Embody, G. C. 1915. The farm fishpond. Cornell reading courses, 4:313–352. pls. 4.

Embody, G. C. and Gordon, M. 1924. A comparative study of natural and artificial foods of brook trout. Amer. Fish. Soc. Trans. 84:185–200.

Eyferth, B. von. 1909. Einfachste Lebensformen. Braunschweig. pp. 584. pls. 16.

Forbes, S. A. 1887. The lake as a Microcosm. Peoria Sci. Assoc. Bull. pp. 15.

Forbes, S. A. 1893. A preliminary report on the aquatic invertebrate fauna of the Yellowstone National Park, Wyoming, and the Flathead Region of Montana. U. S. Fish Com. Bull., 11:207–258, pls. 14.

Forbes, S. A. 1878. The food of Illinois fishes. Ill. State Lab. Nat. Hist. Bull. 1:71–89. pls. 14.

Forbes, S. A. 1914. Fresh water fishes and their ecology. Urbana. 19 pp. 10 pls.

Forbes, S. A. and Richardson, R. E. 1919. The Fishes of Illinois. 2d ed. 3 vols. and atlas.

Forbes, W. T. M. 1910. The aquatic caterpillars of Lake Quinsigamond. Psyche, 17:219–227. 1 pl.

Forbes and Richardson. 1913. Studies in the biology of the upper Illinois River. Ill. State Lab. Nat. Hist. Bull. 20:482–574. pls. 20.

Forel, F. A. 1892–1904. Le Leman, monographie limnologique. In 3 volumes. Lausanne.

Francé, R. H. 1910. Die Kleinwelt des Süssawassers. Leipzig. pp. 160. pls. 50.

Früh and Schroeter. 1904. Die Moore der Schweiz, mit Berücksichtigung der gesammten Moorfrage.

Gage, S. H. 1893. The lake and brook lampreys of New York. Wilder Quarter-Century Book. Ithaca.

Grout, A. J. 1905. Mosses with a hand lens. 208 pp. N. Y.

Haeckl, Ernst. 1893. Planktonic studies. (Trans. by G. W. Field) Rept. U. S. Commissioner of Fish and Fisheries for 1889–1891, pp. 565–641.

Hancock, J. L. 1911. Nature sketches in temperate America. pp. 451. Chicago.

Hankinson, T. L. 1908. Biological Survey of Walnut Lake. Rept. Mich. Geol. Surv., pp. 157–271.

Harris, J. A. 1903. An ecological catalogue of the crayfishes belonging to the genus Cambarus. Kansas Univ. Science Bull., 2:51–187.

Hart, C. A. 1895. On the entomology of the Illinois River. Ill. State Lab. Nat. Hist. Bull. 4:140–237, pls. 4.

Headlee, T. J. 1906. Ecological notes on the mussels of Winona, Pike and Center Lakes of Kosciusko County, Indiana. Biol. Bull. 11:305–316.

Hentschel, E. 1909. Das Leben des Süsswassers. Munich. pp. 336.

Herms, W. B. 1907. An ecological and experimental study of the Sarcophagidæ with relation to lake beach débris. Jour. Exp. Zool., 4:45–83.

Herrick, C. L. and Turner, C. H. 1895. Synopsis of the entomostraca of Minnesota. Geol. and Nat. Hist. Survey, Minn., Zool. Series II. pp. 337. pls. 81.

Howard, A. D. 1914. Experiments in propagation of fresh-water mussels of the Quadrula group. U. S. Bureau Fish. Document No. 801. pp. 52. pls. 6.

Howard, A. D. 1915. Some exceptional cases of breeding among the Unionidæ. The Nautilus. 29:4–11.

Hudson, G. T. and Gosse, P. H. 1889. The Rotifera or wheel-animalcules. Vol. I. pp. 128. 15 pls. Vol. II. pp. 144. 30 pls.

Hungerford, H. B. 1919. The biology and ecology of aquatic and semi-aquatic Hemiptera. Kansas Univ. Sci. Bull. 11:1–265, 29 pls.

Jennings, H. S. 1900. The Rotatoria of the United States. U. S. Fish Comm. Bull. 20:67–104.

Jordan and Evermann. 1904. American food and game fishes. New York.

Juday, C. 1897. The plankton of Turkey Lake. Ind. Acad. Sci. for 1896 Proc. pp. 287–296. 1 map.

Juday, C. 1904. Diurnal movement of plankton crustacea. Wisc. Acad. Sci. Trans. 14:524–568.

Juday, C. 1907. Studies on some lakes in the Rocky and Sierra Mountains. Wisc. Acad. Sci. Trans. 15:781–794. 2 pls. and 1 map.

Juday, C. 1908. Some aquatic invertebrates that live under anaerobic conditions. Wisc. Acad. Sci. Trans. vol. XVI, Part I.

Juday, C. 1915. Limnological studies of some lakes in Central America. Wisc. Acad. Sci. Trans. 18:214–250.

Juday, C. 1916. Limnological Apparatus. Wisc. Acad. Sci. Trans. 18:566–592.

Juday, C. 1926. A third report on limnological apparatus. Wisc. Acad. Sci. Trans. 22:299–314.

Kellogg, V. L. The net-winged midges of North America. Calif. Ac. Sci. Proc. (3) 3:187–223. 5 pls.

Kent, W. S. 1880–1882. A manual of the infusoria. pp. 913: pls. 51.

Kerner and **Oliver.** 1902. Natural history of plants. 2 vols. pp. 1760. London.

Kofoid, C. A. 1903 and 1908. The plankton of the Illinois River. Ill. State Lab. Nat. Hist. Bull. 6:95–629. 1908, 8:1–361.

Knauthe, Karl. 1907. Das Süsswasser. Neudamm. pp. 663.

Lampert, K. 1910. Das Leben der Binnengewässer. Leipzig. pp. 856. pls. 17. 2d edit.

Lefevre and **Curtis.** 1910. Reproduction and parasitism in the Unionidæ. Jour. Exp. Zool. 9:79–115. pls. 5.

Leidy, Joseph. 1870. Fresh-water Rhizopods of North America. U. S. Geol. Survey Rept. Vol. XII.

Lloyd, J. T. 1914. Lepidopterous larvæ from swift streams. N. Y. Ent. Soc. Jour. 22:145–152. 2 pls.

Lloyd, J. T. 1921. North American caddisfly larvae. Lloyd Library Bull. 1:1–24.

Lloyd, J. T. 1915. Notes on Brachycentrus nigrisoma. Pomona Jour. Ent. and Zool. 7:81–86, 1 pl.

Lloyd, J. U. 1882. Precipitates in fluid extracts. Am. Pharmaceut. Assn. Proc. 30:509–518. (Also 1884, 32:410–419).

Lorenz, J. L. 1898. Der Hallstätter See. Geogr. Ges. Wien. Mitt. Bd. 41.

Lundbeck, J. 1926. Die Bodentierwelt Norddeutscher Seen. Arch. Hydrobiol. Suppl. 7, 1:1–160, 5 pl., 19 fig.

Lutz, F. 1913. Factors in aquatic environment. N. Y. Ent. Soc. Jour. 27:1–4.

MacGillivray, A. D. 1903. Aquatic Chrysomelidæ. N. Y. State Mus. Bull. 68:288–312.

Marsh, C. D. 1903. The plankton of Lake Winnebago and Green Lake. Wisc. Geol. and Nat. Hist. Surv. Bull. No. 12, Sci. Ser. 3.

Martynov, A. 1927. Contributions to the aquatic entomofauna of Turkestan. USSR Acad. Sci. Mus. Zool. Ann. 28:162–193.

Matheson, R. 1912. The Haliplidæ of North America north of Mexico. N. Y. Ent. Soc. Jour. 20:157–193.

Matheson, R. 1929. A handbook of the mosquitos of North America. Springfield, 209 pp.

Meister, Fr. 1912. Die Kieselalgen der Schweiz. pp. 254. pls. 48. Bern.

Miall, L. C. 1895. The natural history of aquatic insects. pp. 395. London.

Miall and **Hammond.** 1900. The Harlequin fly (Chironomus) 196 pp. Oxford.

Moore, Emmeline. 1915. The Potamogetons in relation to pond culture. U. S. Bur. Fisheries Bull. 33:251–291. 17 pls.

Moore, Emmeline. 1926–9. (Director) Biological Surveys of New York State river systems: 1926, the Genesee River; 1927, the Oswego River; 1928, the Erie-Niagara drainage; 1929, the Champlain drainage.

Murphy, Helen E. 1922. Notes on the Biology of some North American species of mayflies. Lloyd Library Bull. 22:1–46.

Muttkowski, R. A. 1918. The fauna of Lake Mendota—a qualitative and quantitative survey with special reference to insects. Wisc. Acad. Sci. Trans. 19:374–482.

Nachtrieb, H. F. 1912. The leeches of Minnesota. Geol. and Nat. Hist. Survey of Minn., Zool. Series No. V.

Naumann, F. 1918. Ueber die naturliche Nahrung des limnischen Zooplanktons. Lunds Univ. Arsk. (n.s.) 14:1–48.

Naumann, F. 1921–23. Spezielle Untersuchungen ueber die Ernahrungsbiologie des thierischen Limnoplanktons. Lunds Univ. Arsk. (n.s.) 4: 3–27 and 6:2–17.

Needham and Betten. 1901. Aquatic insects in the Adirondacks. N. Y. State Mus. Bull. 47.

Needham and Hart. 1901. The dragonflies (Odonata) of Illinois, with descriptions of the immature stages. Ill. State Lab. Nat. Hist. Bull. 6: 1–94. pl. 1.

Needham and Williamson. 1907. Observations on the Natural History of diving beetles. Amer. Nat. 41:477–494.

Needham, J. G. 1908. Report of the Entomologic Field Station conducted at Old Forge, N. Y., in the summer of 1905. N. Y. State Mus. Bull. 124:156–248.

Needham, J. G. The burrowing Mayflies of our larger lakes and streams. U. S. Bur. Fish. Bull. 26:269–292, 12 pls.

Needham, J. G. and Heywood, H. B. 1929. A handbook of the Dragonflies of North America. 378 pp. Springfield, Ill.

Needham, J. G. and Claassen, P. W. 1925. A Monograph of the Plecoptera of North America North of Mexico. Thomas Say Foundation Monographs, 2:1–397.

Needham, J. G. and Needham, P. R. 1930. A guide to the study of Freshwater biology. Revised edition, 96 pp., Springfield, Ill.

Needham, P R. 1928. A net for the capture of stream drift organisms. Ecology, 9:339–342.

Noyes, Alice A 1914. The biology of the net-spinning Trichoptera of Cascadilla Creek. Ann. Ent. Soc. Am. 7:251–272. 2 pls.

Ortmann, A. E. 1907. The crawfishes of the state of Pennsylvania. Mem. Carnegie Mus. Pittsburgh, 2:343–523.

Osburn, R. C. 1903. The Adaptation to Aquatic Habits in Mammals. Amer. Nat. 37:651–665.

Parson, H. deB. 1888. The displacement and the area curves of fish. Trans. Amer. Soc. Mech. Engineers. 9:679–695.

Pearse, A. S. 1910. The crawfishes of Michigan. Mich. State Biol. Surv. Rept. 1:9–22.

Pearse, A. S. 1915. On the food of the small shore fishes of the waters near Madison, Wisconsin. Wisc. Nat. Hist. Soc. Bull. 13:7–22.

Pearse, A. S. 1921. The distribution and food of the fishes of three Wisconsin lakes in summer. Univ. of Wisc. Studies 3:1–61.

Paulmier, F. C. 1905. Higher Crustacea of New York City. N. Y. State Mus. Bull. 91. pp. 78.

Platt, Emilie L. 1916. The population of the "blanket-algæ" of fresh water pools. Amer. Nat. 49:752–762.

Reamur, M. dê. 1734–1742. Mémoires pour servir a l'histoire des insectes. 7 vols. Paris.

Reed, H. D. and Wright, H. A. 1909. The Vertebrates of the Cayuga Lake Basin. Am. Philos. Soc. Proc. 48:370–459.

Reese, A. M. 1915. The alligator and its allies. pp. 341. pls. 28. N. Y. and London.

Richardson, H. 1905. A monograph of the Isopods of North America. U. S. Nat. Mus. Bull. 54:727.

Richardson, R. E. 1921. The small bottom and shore fauna of the middle and lower Illinois River and its connecting lakes, etc. Nat. Hist. Surv. Ill. Bull. 13:363–522, with charts.

Richmond, E. A. 1920. Studies on the biology of aquatic Hydrophilidae. Amer. Mus. Nat. Hist. Bull. 42:1–94, 16 pls.

Reighard, J. 1894. A biological Examination of Lake St. Clair. Bull. Mich. Fish Comm., No. 4. 60 pp. 2 pls. and 1 map.

Reighard, J. 1910. Methods of studying the habits of fishes, with an account of the breeding habits of the horned dace. U. S. Bur. Fish. Bull. 28:1112–1136.

Rösel von Rosenhof, August Johann. 1846–1861. Insecten Belustigung. 4 vols.

Russel, I. C. 1895. Lakes of North America. pp. 125. pls. 23. Boston.

Russel, I. C. 1898. Rivers of North America. pp. 327. pls. 17. New York.

Scott, Will. 1911. The fauna of a solution pond. Indiana Univ. Studies. 48 pp.

Scourfield, D. J. 1896. Entomostraca and the surface film of water. Jour. Linn. Soc. Zool. 25:1–9. pls. 2.

Scourfield, D. G. 1900. Notes on Scapholeberis mucronatus and the surface film of water. Jour. Queckett Micr. Club. (2) 7:309–312.

Sellards, E. H. 1914. Some Florida lakes and lake basins. 6th annual Rept. Fla. State Geol. Survey. pp. 115–160.

Shantz, H. L. 1907. A biological study of the lakes of the Pike's Peak Region. Amer. Micr. Soc. Trans. pp. 75–98. pls. 2.

Sherff, E. E. 1912. The vegetation of Skokie Marsh. Bot. Gaz., 54:415–435.

Shelford, V. E. 1913. Animal communities in Temperate America. Chicago. pp. 362.

Shelford, V. F. 1911–13 Ecological succession. Biol. Bull. 21:127–151 and 22:1–38 and 23:59–99.

Shelford, V. F. 1929. Laboratory and field ecology. Baltimore. 608 pp.

Sibley, C. K. and other members of the scientific staff of Cornell University. 1926. A preliminary biological survey of the Lloyd-Cornell Reservation (near McLean, N. Y.). Lloyd Libr. Bull. 27:1–246.

Smith, G. M. 1920. The phyto-plankton of the inland lakes of Wisconsin, Part 1. Wisc. Geol. Nat. Hist. Surv. Bull. 57:1–243, 51 pls.

Snow, Julia W. 1902. The Plankton algæ of Lake Erie. Bull. U. S. Fish. Comm. 22:371.

Steuer, A. 1910. Planktonkunde. pp. 723. Leipzig.

Stokes, A. C. 1888. A preliminary contribution toward a history of the fresh-water Infusoria of the United States. Trenton. Nat. His. Soc. Jour. Vol. I. No. 3.

Stokes, A. C. 1896. Aquatic microscopy for beginners. pp. 326. 3d edit. Trenton.

Strodtman. 1898. Ueber die vermeintliche Schädlichkeit der wasserblüthe Biol. Sta. plön. Forschungsber. 6:206–212.

Surber, T. 1912. Identification of the glochidia of freshwater mussels. U. S. Bureau Fish. Document No. 771. pp. 10. 3 pls. Also 1915 Document No. 813. pp. 9. 1 pl.

Swammerdam, Johannes. 1685. Historia insectorum generalis. pp. 212. pls. 13.

Tilden, Josephine. 1910. Minnesota algæ. Report of Minn. Surv., Bot. Series VIII. Vol. I. pp. 328. pls. 20.

Transeau, E. N. 1906. The bogs and bog flora of the Huron River Valley. Bot. Gaz., 40:351–428.

Transeau, E. N. 1908. The relation of plant societies to evaporation. Bot. Gaz., 45:317–331.

Vorhies, C. T. 1909. Studies on the Trichoptera of Wisconsin. Wisc. Acad. Sci. Trans. 16:647–738. 10 pls.

Walton, L. B. 1915. Euglenoidina, Ohio State Univ. Bull. Vol. XIX. No. 5.

Ward, H. B. 1896. A biological examination of Lake Michigan in the Traverse Bay region. Mich. Fish. Com. Bull. No. 6. pp. 100. pls. 5.

Ward and **Whipple.** Editors. American fresh-water biology. Chapters by many specialists. New York. 1111 pp.

Warming, E. 1909. Ecology of plants. An introduction to the study of plant communities. Oxford. (Transl. by Percy Groom.)

Weckel, Ada L. 1907. The fresh-water Amphipoda of North America. U. S. Nat. Mus. Proc., 32:25–58.

Weckel, Ada L. 1914. Free-swimming fresh-water Entomostraca of North America. Amer. Mic. Soc. Trans. 33:165–203.

Welch, Paul S. 1927. Limnological investigations on northern Michigan lakes. I. Physical-chemical studies on Douglas Lake. Mich. Acad. Sci. Arts and Let. 8:421–451. 1 map.

Wesenberg-Lund, C. 1910. Grundzüge der Biologie des Süsswasserplanktons. Internat. Rev. Hydrobiol. und Hydrog. Biol. Suppl. I. (zu Bd. III.)

Wesenberg-Lund, C. 1923. Contributions to the biology of the Rotifera. Copenhagen.

West, G. S. 1904. A treatise on the British freshwater algae. pp. 372. Cambridge.

West, W. and **West, G. S.** 1904–1908. A monograph of the British Desmidiaceæ. Ray Society publications. Vol. I. pp. 224. pls. 32. Vol. II. pp. 204. pls. 64. Vol. III. pp. 273. pls. 95.

Whipple, George C. 1914. The microscopy of drinking water. New York and London. 3d. edition. pp. 409, pls. 19.

Wilson, C. B. 1923. Water beetles in relation to pond fish culture. U. S. Bur. Fish. Bull. 39:231–343.

Wolcott, R. H. 1905. A review of the genera of the water mites. Am. Micro. Soc. Trans. pp. 161–243.

Wolle, Francis. 1887. Fresh-water algæ of the United States. Vol. I. pp. 364: Vol. II. pls. 210.

Wolle, Francis. 1892. Desmids of the United States. pp. 182. pls. 64.

Wolle, Francis. 1894. Diatomaceæ of North America. pp. 45. pls. 112.

Wright, A. H. 1914. North American Anura: life-histories of the Anura of Ithaca, N. Y. Carnegie Inst. Pub. No. 197. pp. 98. pls. 21.

Zacharias, O. 1891. Die Tiere- und Pflanzenwelt des Süsswassers. Leipzig.

Zacharias, O. 1907. Das Süsswasserplankton. Leipzig.

LIST OF INITIALS AND TAIL-PIECES

INDEX